Effective Cross-Border Monitoring Systems for Waterborne Microbial Pathogens

Effective Cross-Border Monitoring Systems for Waterborne Microbial Pathogens

A Plan for Action

Edited by

Joan. B. Rose and Erin. A. Dreelin

Publishing

Published by IWA Publishing, Alliance House, 12 Caxton Street, London SW1H 0QS, UK
Telephone: +44 (0) 20 7654 5500; Fax: +44 (0) 20 7654 5555; Email: publications@iwap.co.uk
Web: www.iwapublishing.com

First published 2008

Cover design by www.designforpublishing.co.uk
Index provided by Christine Graham

British Library Cataloguing in Publication Data
A CIP catalogue record for this book is available from the British Library

Library of Congress Cataloging- in-Publication Data
A catalog record for this book is available from the Library of Congress

ISBN: 1843391694
ISBN13: 9781843391692

Contents

Preface

J. B. Rose and E. A. Dreelin

Water is the basis of all life on our planet. Throughout the history of humankind, the development of communities along coastlines, where food and water were plentiful, meant waterways were also used for travel, trade, industrial, and municipal purposes. As we moved from a society that relied on hunting and gathering to a societal structure based on farming, the connection between water and food supply became intertwined. Early irrigation schemes along the Nile meant communities could grow and support themselves and supply others via trade. In other areas, the reliance on "nature" meant that the types of crops and amounts of food that could be produced were restricted, until water systems were established.

As science and knowledge progressed, we learned more about how water moved in a cycle - how rain, snow, rivers, coasts, lakes and groundwater are connected. Engineering paved the way for further connecting waters by moving, storing, accessing and treating waters. With these advancements communities flourished.

 Effective Cross-Border Monitoring Systems for Waterborne Microbial Pathogens. Edited by J. B. Rose and E. A Dreelin. ISBN: 9781843391692. Published by IWA Publishing, London, UK.

Beyond being purely utilitarian, for many of us there is a spiritual side to water. The beauty of the natural world that water has provided includes the vast array of plants, animals, and physical features integral to the waters of the world and their ecosystems. Science has begun to unravel these ecological water-based structures/communities and understand what contributes to the health or demise of these systems, and how these systems effect our well being.

Scientists continue to investigate contamination of water and waterborne disease. Early civilizations realized that water is the great dissolver and that it carries away wastes. As communities grew, sewage increased and water pollution became a prevalent problem, particularly around large cities and populated areas. By the 1800s, we learned that water could transmit disease and pollution could kill life in a river. As a result, humans used dilution or treatment technologies to reduce the threat of waterborne disease. But despite these efforts, water pollution remains the single greatest cause of mortality in the world. In addition, the Asian tsunami of 2004 and Gulf of Mexico Hurricane Katrina of 2005, were dramatic examples of the power of water. Storm events can cause flooding and damage to life and property. Flood control has long been a goal of humans striving to control the water environment. Flooding is the single most important natural disaster associated with both direct and indirect effects on human health.

Science has progressed our understanding of the complexities of natural hydrologic systems. But we currently have an engineered water cycle associated with advancements in water science and technology. Investigation, information, knowledge, education, and communication of this water science must continue. Communication between scientists, decision-makers and the public will improve the decisions that society makes regarding the management and protection of such a vital resource.

The societal benefits for managing our waters (both in terms of quantity and quality) should be clear. Yet conflicts between people, communities, states, and nations have arisen regarding the sharing of waters and protecting water quality, quantity and human uses. This has been highlighted in particular in those watersheds and basins which span political boundaries.

To date, transboundary governmental agreements to protect the quality of waters are rare. Early in history, the focus of water quality protection was prevention of acute disease such as preventing typhoid. We have relied heavily on "bacterial indicators" and treatment technologies to reduce the threats of microbial water pollution. In the last 40 years, attention has focused on the chemical quality of water. Yet microbial water quality remains the greatest concern globally. Advances in the field of microbial monitoring will provide information critical to decisions that governments will need to make collectively to sustain water quality and protect health. It is our hope that this book will provide historical significance, address societal considerations, and introduce key tools and frameworks that will have application to large water basins and lead toward water quality agreements that diminish the risk of waterborne disease to communities in the future.

We would like to thank the authors of each of the chapters. We would also like to thank the following reviewers: Joseph Cotruvo, Mary Culver, Mary Durfee, Pierre Payment, and Melita Stevens. S. Sattar and J. Tetro wish to thank Ms. Susan Springthorpe of CREM for her valuable comments on their chapter. We would like to thank Audrey Tan and Rachel McNinch for their assistance with formatting the book. We wish to thank Dr. Jiaguo Qi and the Center for Global Change and Earth Observations (http://www.globalchange.msu.edu/index.html) for the images appearing on the cover and at the beginning of each chapter.

1

Transboundary Monitoring of International Waters: Critical Questions for Microbial Water Quality

J. B. Rose, E.A. Dreelin, and S. Weisberg

 Effective Cross-Border Monitoring Systems for Waterborne Microbial Pathogens. Edited by J. B. Rose and E. A. Dreelin. ISBN: 9781843391692. Published by IWA Publishing, London, UK.

The Aral Sea. Image: MSU Center for Global Change and Earth Observations. At the beginning of each chapter, we provide an image of a transboundary waterbody. While this book uses the Great Lakes as a case study, the issues discussed are not unique to the Great Lakes and may be applicable to other regions around the world.

1.1 INTRODUCTION

Across the globe, there are 20 international lakes and 261 international river basins, which are also called watersheds in the US or catchments in Europe. These water resources are shared across political boundaries and include 71 basins in Europe, 60 in Africa, 53 in Asia, and 39 and 38 in North and South America, respectively. All of these basins are shared by two or more countries.

Wolf et al. (1999) have done a superb job in summarizing and updating the international river basins of the world, updated with changing political boundaries http://www.transboundarywaters.orst.edu/publications/register/ regi ster_paper.html#). Some of the largest international river basins are the Niger, Nile, and Zaire in Africa; the Tigris-Euphrates/Shatt al Arab, Indus, Mekong, Ob in Asia; the Danube and Rhine in Europe; and the Amazon in South America (Table 1.1). Flow from these rivers is equal to just below 50% of the total world's water runoff (total of 19, 200 km^3 per day, Shiklomanov, 2000). The largest international lakes are found in Africa, Asia and North America; these include Lake Chad, Lake Victoria, the Caspian Sea and the Great Lakes (Table 1.2). While there are a number of international lakes in Europe, they are all relatively small in size and typically involve two countries.

Table 1.1 Examples of some of the largest international water basins of the world (Wolf et al. 1999)

Continent	**AFRICA 60 International River Basins**
Basin	**Countries**
Congo/Zaire Basin area: 3,691,000 km^2	Congo, Democratic Republic of (Kinshasa) Central African Republic Angola Congo, Republic of the (Brazzaville) Zambia Tanzania, United Republic of Cameroon Burundi Rwanda Sudan Gabon Malawi Uganda
Lake Chad Basin area: 2,388,700 km^2	Chad Niger Central African Republic Nigeria Algeria Sudan Cameroon Chad, claimed by Libya Libya

Table 1.1 continued. Examples of some of the largest international water basins of the World (Wolf et al. 1999)

Continent	**AFRICA 60 International River Basins**
Basin	**Countries**
Niger Basin area: 2,113,200 km^2	Nigeria Mali Niger Algeria Guinea Cameroon Burkina Faso Benin Ivory Coast Chad Sierra Leone
Nile Basin area: 3,031,700 km^2	Sudan Ethiopia Egypt Uganda Tanzania, United Republic of Kenya Congo, Democratic Republic of (Kinshasa) Rwanda Burundi Egypt, administered by Sudan Eritrea Sudan, administered by Egypt Central African Republic
Zambezi Basin area: 1,385,300 km^2	Zambia Angola Zimbabwe Mozambique Malawi Tanzania, United Republic of Botswana Namibia Congo, Democratic Republic of (Kinshasa)

Table 1.1 continued. Examples of some of the largest international water basins of the world (Wolf et al. 1999)

Continent	**ASIA 53 International River Basins**
Basin	**Countries**
Aral Sea Basin area: 1,231,400 km^2	Kazakhstan Uzbekistan Tajikistan Kyrgyzstan Afghanistan Turkmenistan China Pakistan
Ganges- Brahmaputra- Meghna Basin area: 1,634,900 km^2	India China Nepal Bangladesh India, claimed by China Bhutan India control, claimed by China Myanmar (Burma)
Indus Basin area: 1,138,800 km^2	Pakistan India China Afghanistan Chinese control, claimed by India Indian control, claimed by China Nepal
Mekong Basin area: 787,800 km^2	Laos, (People's Democratic Republic of) Thailand China Cambodia (Kampuchea) Vietnam Myanmar (Burma)
Ob Basin area: 2,950,800 km^2	Russia Kazakhstan China Mongolia

Table 1.1 continued. Examples of some of the largest international water basins of the world (Wolf et al. 1999)

Continent	ASIA 53 International River Basins
Basin	**Countries**
Tarim Basin area: 2,950,800 km^2	China Chinese control, claimed by India Kyrgyzstan Tajikistan Pakistan Afghanistan
Tigris- Euphrates/Shatt al Arab Basin area: 789,000 km^2	Iraq Turkey Iran Syria Jordan Saudi Arabia

Continent	EUROPE 71 International River Basins
Basin	**Countries**
Danube Basin area: 790,100 km^2	Romania Hungary Austria Yugoslavia (Serbia and Montenegro) Germany Slovakia Bulgaria Bosnia and Herzegovina Croatia Ukraine Czech Republic Slovenia Moldova Switzerland Italy Poland Albania

Table 1.1 continued. Examples of some of the largest international water basins of the world (Wolf et al. 1999)

Continent	**EUROPE 71 International River Basins**
Basin	**Countries**
Rhine Basin area: 172,900 km^2	Germany Switzerland France Belgium Netherlands Luxembourg Austria Liechtenstein Italy
Vistula/Wista Basin area: 194,000 km^2	Poland Ukraine Byelarus Slovakia Czech Republic
Volga Basin area: 1,554,900 km^2	Russia Kazakhstan Byelarus
Continent	**NORTH AMERICA 39 International River Basins**
Basin	**Countries**
Colorado Basin area: 172,900 km^2	United States of America Mexico
Columbia Basin area: 668,400 km^2	United States of America Canada

Table 1.1 continued. Examples of some of the largest international water basins of the world (Wolf et al. 1999)

Continent	**NORTH AMERICA 39 International River Basins**
Basin	**Countries**
Great Lakes	United States of America Canada
Mississippi Basin area: 3,226,300 km^2	United States of America Canada
Rio Grande Basin area: 656,100 km^2	United States of America Mexico
Continent	**SOUTH AMERICA 38 International River Basins**
Basin	**Countries**
Amazon Basin area: 5,883,400 km^2	Brazil Peru Bolivia Colombia Ecuador Venezuela Guyana Suriname French Guiana
La Plata Basin area: 2,954,500 km^2	Brazil Argentina Paraguay Bolivia Uruguay
Lake Titicaca-Poopo System Basin area: 111,800 km^2	Bolivia Peru Chile
Orinoco Basin area: 927,400 km^2	Venezuela Colombia Brazil

Table 1.2 Examples of Some of the Largest International Lakes in the World

Continent	**AFRICA International Lake Basins**
Basin	**Countries**
Chad Lake 25,760 km^2 1,035,000 km^2 catchment 22 million people	Chad Niger Nigeria
Victoria Lake 69,485 km^2 Nile river basin, 193,000 km^2 catchment 20 million people	Tanzania Uganda
Nyasa Lake 30,044 km^2 100,500 km^2 catchment	Malawi Mozambique Tanzania
Tanganyika Lake 32,893 km^2 26,300,000 km^2 Catchment	Tanzania, Zaire, Zambia Burundi
Continent	**ASIA International Lake Basins**
Lake	**Countries**
Caspian Sea Lake 394,299 km^2 1,400,000 km^2 Catchment	Azerbaijan Russia Kazakhstan Turkmenistan Iran[1]
Aral Lake 33,800 km^2 1,549,000 km^2 catchment	Kazakhstan Uzbekistan
Lake Peipsi Lake 3,429 km^2 47,800 km^2 catchment 1 million people	Estonia Russia

Table 1.2 continued. Examples of Some of the Largest International Lakes in the World

Continent	**EUROPE International Lake Basins**
Lake	**Countries**
Lake Bodensee /Constance Lake 572 km^2 11,488 km^2 Catchment	Austria Germany Switzerland
Continent	**NORTH AMERICA International Lake Basins**
Basin	**Countries**
Great Lakes: Lakes 244,160 km^2 (Superior, Michigan, Ontario, Huron, and Erie) 521,830 km^2 catchment 40 million people	Canada United States
Continent	**SOUTH AMERICA International Lake Basins**
Basin	**Countries**
Titicaca Lake 8,135 km^2 56,270 km^2 Catchment 2 million people	Bolivia Peru

Perhaps it is the fragmentation of watersheds by states and nations that has led to poor investment in protection of the water quality. A global assessment of water quality has not been undertaken and one can only assume a significant level of contamination exists based on the numbers of waterborne illnesses and the lack of sewage treatment throughout the world. Historically, humans have used the "best quality" water for drinking water supply, thus protection efforts are geared toward drinking water supplies, with little concern shown for receiving water basins and downstream users.

More attention generally has been paid to lakes than river basins. Notable drops in water availability due to water withdrawals and climate change are described for many of the lakes, including Lake Chad and the Aral Sea. Problems with fishing stocks, eutrophication and algal blooms are commonly observed in many lakes across the world. The United Nations has funded a number of programs that address the status of lakes around the world, including in some cases water quality (www.worldlakes.org; http://www.ilec.or.jp/ database /database.html). The water quality variables included most often in monitoring programs were dissolved oxygen, biological oxygen demand, and nutrients such as phosphorus and nitrogen (including nitrite and nitrate). However, very few monitoring programs included any microbiological monitoring. In a few cases in Europe only coliform bacteria were monitored and reported. It is clear that what is needed is a comprehensive database on microbial water quality, trends, and causes of water pollution.

Waterborne diseases include those diseases that historically caused widespread outbreaks and severe health effects such as cholera, typhoid and poliovirus. All of these diseases are associated with exposure to human fecal waste. While these diseases have been eliminated in the developed world, they remain a significant problem globally. Recently, new infectious diseases and outbreaks have been increasing around the world at unprecedented rates (WHO 2007). This increase may be due in part to antibiotic resistance, genetic emergence of new strains of pathogens, and outbreaks of zoonotic diseases (spread from animals to humans) thus humans are a greater risk from polluted water than ever before. By 1970, 39 new diseases had been identified and since 2002 more than 1,100 epidemics have been documented including cholera, polio and *Cryptosporidium*. Throughout the world, cholera and typhoid are associated with water contamination (Table 1.3) and can be used as an index. While cholera is typically reported as directed by the Global Health Assembly, typhoid is not and thus uncertainty about some estimates hamper the true assessment of the water-health index.

Table 1.3 Water Pollution and Health Index (Rose and Farrah 2007)

Continent	# of International River Basins (1999)	Cholera 1995-2005	Typhoid Estimates Per year***	Percentage with access to sanitation (2002) ****
Africa	60	1,487,041	408,873	55
Asia	53	196,524	10,118,879	56
Europe	71	1,128	19,144	83
North America	39	119*	453*	98
Central and South America	38	199,230**	273,518***	75
Total	261			58

*Estimates for NA include only USA and Canada and South America includes Central America;**64,253 in Central America; ***Crump et al, 2004; ****Calculated from Gleick 2006

Water pollution remains a critical issue that directly influences global health. Water pollution also affects the scarcity of water given that humans require clean, safe water. Water security, early warning, protection and restoration for desired uses will require significant financial investments. In order to achieve sustainable access to safe water (both in terms of sufficient quantity and quality), local, regional, national and international cooperation will be required. Water shortages will likely impact 2.7 billion people in about 20 years, threatening the global food supply and the economies of the more than 50 countries where water will become scarce (www.worldwatercouncil.org; Gleick, 1993; Shmueli, 1999). Currently, water quality is not adequately addressed in most international water treaties (Shmueli, 1999). However, one example that could be used as a model for other international basins is the Great Lakes Water Quality Agreement between Canada and the US. Even with this agreement in place, discussion continues on how to better include pathogens and improve assessment of microbial water quality of the ground, surface and coastal systems in the Great Lakes.

1.2 THE GREAT LAKES BASIN

The Great Lakes comprise the largest source of fresh water on the surface of the Earth, holding nearly 84% of freshwater in North America and 20% of the fresh water on the planet (EPA GLNPO 2005). The Great Lakes basin includes Lakes Erie, Huron, Michigan, Ontario and Superior and 5 major connecting rivers including the Detroit River, Niagara River, St. Clair River, St. Lawrence River and St. Mary's River (Figure 1.1; GLIN 2006a). Waters of the Great Lakes support many economically important activities including shipping, drinking water, swimming and fishing. The recreational fishery alone generates approximately $5 billion (USD) each year and the lakes provide critical ecosystems and habitat for many valued species.

Although human diseases such as typhoid and cholera are no longer major problems in the Great Lakes basin, waterborne microbial pathogens still warrant considerable attention. Recent drinking water outbreaks in South Bass Island (Lake Erie, Ohio, 2004), Walkerton, Ontario, (2000) and the largest documented outbreak in North America in Milwaukee, WI (1993) illustrate that this water basin is highly vulnerable to contamination and waterborne microbial pathogens still pose a major threat to human health. Most of these pathogens are transmitted by the fecal-to-oral route in which people are exposed to the pathogens when they ingest or come into contact with drinking or recreational water contaminated with human or animal feces. Although attention and management efforts have focused on treating water and wastewater to prevent microbial contamination, human activities, population growth, emerging pathogens, and poor infrastructure continue to contribute to contamination of surface and ground water with microbial pathogens. This is exacerbated by shifts in weather and climate.

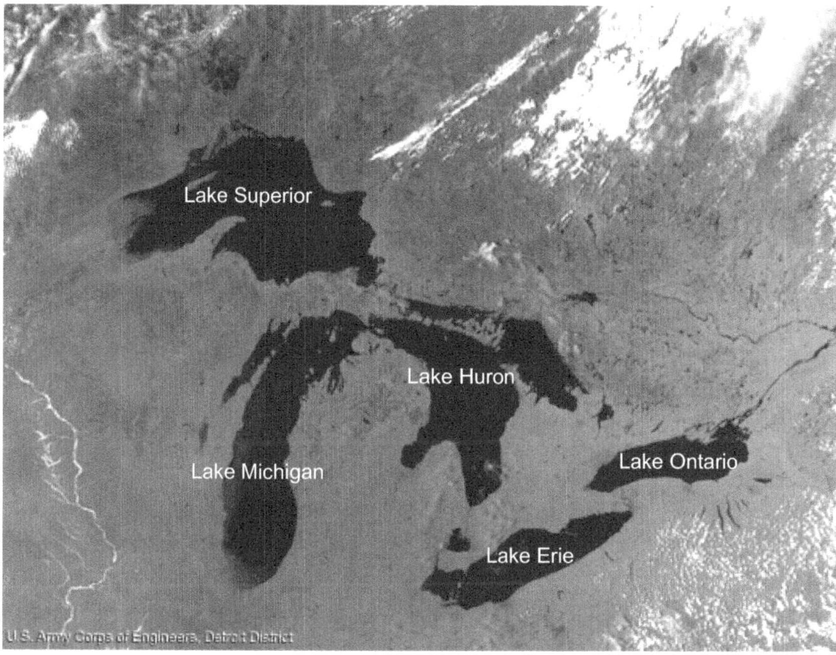

Figure 1.1The Great Lakes (GLIN 2006b).

In order to fully protect water quality and the health of transboundary waters, common goals, databases and management approaches are needed. Water quality agreements should address management of microbial contamination, and must move beyond the indicator system which is an inadequate measure of risk and leads to poor assessment of needs. Therefore, a system whereby pathogens can be monitored directly in the environment and scientific information can be developed to understand their sources, fate and transport is needed.

In the Great Lakes basin, both the US and Canada have worked together to address and control pollution for more that 100 years (see Chapter 2). In the early 1900s, the International Joint Commission (IJC) supported the largest bacteriological monitoring study ever developed to examine cross-boundary pollution in response to typhoid outbreaks (see Chapter 3). This study was revolutionary in its day and can be used to inform current monitoring programs. The societal changes (travel and tourism) and demographic changes have created new susceptibility to disease (see Chapter 4). Emerging pathogens have caused outbreaks in the basin, such as *Cryptosporidium* in Milwaukee, Wisconsin, *E.coli* 0157H7 in Walkerton, Ontario and Norovirus and *Campylobacter* in South Bass Island, Lake Erie Ohio. However, new technologies and approaches (from DNA to satellite imaging) are developing which allow determination of the presence of pathogens and provide information on their fate in the environment. The Global Ocean Observing System (GOOS) is being established world wide and can be used to track harmful algal blooms (see Chapter 5) as well as set up a network of sites for sampling. This system would provide managers with real-time information on rainfall, runoff and contamination and can be used in determining impacts on coastlines and beaches. Finally, Hazard Analysis and Critical Control Point (HACCP) planning provides a framework for developing a comprehensive plan for monitoring and managing microbial contamination in the Great Lakes basin (see Chapter 6). While the World Health Organization moves forward with Water Safety Plans, that incorporate HACCP principals, there is a need to pull together an approach that allows water quality to be adequately assessed in regard to health benefits and prioritizing investments. This means monitoring with a variety of techniques and use of available data at spatial and temporal scales that are appropriate for addressing water quality and health, which can address the global cycle of water (see Chapter 7).

1.3 MICROBIAL PATHOGENS AND THE GREAT LAKES

The Great Lakes basin is home to approximately 30 million people, many of whom drink and recreate in the waters of the basin without any thought to microbial pathogens. However, in the US there has been an increase in the number of disease outbreaks due to waterborne pathogens since the late 1990s. The US Centers for Disease Control (CDC) estimates that there are 2 million infections in the US every year caused by *Giardia* and 300,000 caused by *Cryptosporidium* (CDC 2007). These are but two of many waterborne microbial pathogens that potentially contaminate water. The largest drinking water outbreak in US history occurred in the Great Lakes region (1993 Milwaukee, Wisconsin) where 400,000 people became ill and 100 people died due to the emerging parasite *Cryptosporidium* (MacKenzie et al 1994). Rainfall and untreated sewage as well as animal wastes were to blame for the contamination of Lake Michigan, the water supply for that city. Because *Cryptosporidium* is resistant to chlorination, when filtration at the treatment plant failed to provide enough of a barrier, a massive disease outbreak resulted. Outbreaks have also occurred in Canada, most notably the Walkerton, Ontario outbreak in 2000. Seven people died as a result of *E. coli* O157:H7 contamination of the ground water supply. The water was inadequately disinfected before being distributed to the community for drinking. The water was also contaminated with *Campylobacter.* The source of these pathogens was animal manure that was spread on a nearby farm in close proximity to the well prior to a large storm event, rainfall provided the needed transport mechanism. Five years later, the community is still suffering chronic symptoms. Finally, in 2004, on South Bass Island, Put-In-Bay in Lake Erie, massive groundwater contamination resulted in thousands of illnesses associated with viruses (norovirus) and bacteria (*Campylobacter*) as well as unknown agents causing gastrointestinal illness. The rainfall that year was 150% the 50 year average and a wind event lead to hydraulic mixing of sewage and groundwater (Fong et al. 2007).

In addition to contamination of drinking water, microbial pathogens also pose a risk for contact recreation. In 2006, over 3,000 closings or advisories were posted at US Great Lakes beaches due to the presence of pathogens or indicator organisms, an increase of over 10% from 2005 (NRDC 2007). Globally, the cost of human disease caused by sewage pollution of coastal waters is estimated at 4 million lost 'man-years' annually, which is roughly equivalent to an annual economic loss of approximately $16 billion U.S. For one Lake Michigan beach, net economic losses due to beach closures have been estimated to range from $1,274 to $37,030 per day depending value assumptions used (Rabinovici et al. 2004).

1.4 GREAT LAKES MANAGEMENT FRAMEWORK

1.4.1 Management in the Great Lakes Region

There are many players in the Great Lakes with some involvement in management of water and wildlife resources in the basin. The federal governments of Canada and the United States of America (US), eight US states, two Canadian provinces, four regional institutions, approximately 120 Native American authorities, and thousands of local governments have some legal authority or responsibility for ecosystem management in the basin (Hildebrand et al. 2002). Because multiple players are involved in coordinating and planning, there is often confusion among federal and state officials as to which entity bears ultimate responsibility for management (GAO 2003).

There is a long history of cooperation between the US and Canada over management of the Great Lakes basin. In 1909, the two countries signed the Boundary Waters Treaty. The Treaty commits the two countries to protecting and improving boundary waters of the Great Lakes. The Treaty also created the International Joint Commission (IJC), a binational agency charged with resolving disputes between the two countries (see Chapter 2).

The Great Lakes Water Quality Agreement of 1972 was a response to continuing pollution problems in the Great Lakes (Hildebrand et al. 2002). The goal of the Water Quality Agreement was to "restore and maintain the chemical, physical and biological integrity of the waters of the Great Lakes ecosystem." The Water Quality Agreement expanded the role of the IJC to provide technical assistance, coordinate management actions of Canada and the US, and review how well the two countries were implementing the goals of the treaty. The Agreement was amended in 1978, 1983, and 1987; each revision added more focus on achieving water quality objectives. The 1987 Protocol to the Agreement required the two countries to identify Areas of Concern (AOC), areas within the basin that are extremely polluted and likely to impair beneficial uses. The federal governments are required to work with state/provincial and local partners to develop Remedial Action Plans (RAPs) to clean up the AOCs. The IJC produces a biennial report on the progress made under the Agreement. The Water Quality Agreement and Boundary Waters Treaty form the foundation for US and Canada to jointly manage the resources of the Great Lakes (EPA GLNPO 2005). More information on these agreements is provided in Chapter 2.

Several other efforts address management of water and water-related resources in the Great Lakes. In 1955, US and Canada signed the Convention on Great Lakes Fisheries to ensure collaboration and better protection of fisheries. The Convention established the Great Lakes Fisheries Commission (GLFC) to research means to ensure sustainable fisheries, make recommendations to Canada and the US based on research results, and to control the invasive sea lamprey. On the US side, the eight Great Lakes states (Minnesota, Wisconsin, Illinois, Indiana, Michigan, Ohio, Pennsylvania, and New York) signed the Great Lakes Compact in 1955. The purpose of the Compact was to promote integrated planning and management of the basin across the Great Lakes states, to balance competing uses for Great Lakes resources, and to create an intergovernmental agency to promote collaboration. The Compact established the Great Lakes Commission (GLC) to implement the compact on behalf of the states. In 1999, the Canadian provinces of Quebec and Ontario became associate members of the GLC, making the GLC a binational agency (See Chapter 2)

1.4.2 Management Challenges

The many stressors to the system, including ballast water discharges, sewage discharges, and nonpoint source runoff, further complicate management of Great Lakes waters. Many of the effects from these discharges are cumulative over large portions of the lakes and effective management requires consistent protection on both sides of the border.

Developing cross-border management consistency requires a thorough understanding of stressors and effects. Before implementing potentially costly controls, managers must first develop a risk assessment, characterizing magnitude of the problem and whether problems are getting worse. They must next agree on which sources pose the greatest contribution to the problem. They must also agree on what types of problems are best addressed through local controls and which require consistent cross-border application. These latter two items require understanding of transport and fate, so that managers can identify sources that need to be reduced and receptors that need to be protected or restored.

Managers often do not have this requisite information and one of the biggest hindrances is poorly integrated monitoring. The Great Lakes, as with many other large water body systems, has a myriad of monitoring programs. These can be difficult to integrate because they are conducted at different spatial scales by different management entities, and with different statistical designs. Moreover, these programs often employ different sampling and analytical methods, with different levels of quality assurance. Even when the data are collected in similar methods and with a similar commitment to quality, the data systems that house the information are rarely compatible.

Differences among programs are likely to expand, as there are many new measurement technologies on the horizon. The use of molecular and immunological methods is revolutionizing the way we measure biological responses (Noble and Weisberg 2005). These methods, which are replacing culture-based microbiological methods and hand counting/identification for macrobiota, allow more rapid quantification and a greater level of source specificity (Wade et al. 2005, Blanch et al. 2006). These methods also allow measurement of many emerging pathogens (cancer viruses, antibiotic resistance, toxins) that are not practical to quantify with traditional methods (NRC 2004).

Similarly, remote sensing technologies are bridging spatial and temporal data gaps (See Chapter 4). Satellite imagery allows selected water quality parameters to be examined synoptically and with common measurement techniques across the entirety of a water body (Ocean.us 2007). Moored instrumentation allows continuous measurements of selected parameters at selected sites, allowing a better understanding of the complex diel and daily temporal patterns of condition (Boehm et al. 2002). These new continuous monitoring tools provide increased opportunity for warning systems to be based more on prediction than on fecal monitoring (Nevers et al. 2005, Hou et al. 2006).

While these new methods have the potential to greatly expand our knowledge base, they are mostly being implemented on a project-specific basis. There is no clear means for determining when these technologies are ready to transition from research to operational monitoring. There is also no means for determining which specific techniques from within a class of techniques are most appropriate for this transition.

Adding to this problem is that we often do not have an interpretational framework for these new measures. We can measure the spatial and temporal distribution of a parameter, but this is difficult to interpret without an agreed upon threshold of concern. This is particularly problematic for many of the emerging pathogen measures, for which expensive epidemiological studies are necessary to define an interpretational framework.

1.5 CROSS-BORDER ISSUES

Integration of monitoring and management is further complicated when there are cross border issues, as there are on the Laurentian Great Lakes. Borders tend to create barriers because of different perspectives and interests of the parties. Within a border, the focus of monitoring can remain on the common goal of defining the problem and its solution. When borders are involved, legal issues, such as who is financially responsible for the solution, taint such questions. Even if there is agreement on the nature and source of the problem, differing legal frameworks complicate cross-border management, as the procedural steps and time necessary to implement a solution vary among entities.

Differences in technical abilities may also complicate cross border management. Even within a country, there are differing levels of monitoring and assessment sophistication between urban and rural areas. Differences in monitoring investment among entities can be severe, such as along the southern US-Mexico border. There are also differences in approaches between countries that will lead to variation in technologies and techniques employed.

There have been great strides towards harmonization of water management in the Great Lakes. The US BEACH Act has facilitated standardization of beach monitoring for US coastal states including the Great Lakes states. The World Health Organization (WHO) has begun articulating approaches for Water Safety Plans, which address monitoring and control of water quality at the watershed level for drinking water. The U.S. National Academy of Sciences has recently released a report that presents a case for a tiered approach to microbial water quality testing using advanced methodologies and models (NRC 2004).

Pathogens also create particular problems that are very different than chemicals. Single exposures to even small concentrations of some of these pathogens cause acute disease, with devastating consequences and even death. Indicator bacteria are used for identifying the presence fecal contamination but indicators do not provide information on the source of contamination, which is important for determining which types of pathogens might be present and how one might fix the problem. Different yard sticks of safety and different management options are used for microbial contamination (e.g., same system elicits a beach advisory or beach closure). Thus there is no universal acceptable level of pathogen contamination. Complacent attitudes begin to emerge overtime regarding indicators but pathogen detection may initiate a greater concern that may or may not be warranted.

While this book focuses on cross-border issues for the Great Lakes, the issues are not unique to this location. For example, there are noticeable differences in beach water quality between the United States and Mexico on their Pacific border (Noble et al 2003), with concern that beach water quality in the United States is negatively affected by cross-border flow (Gersberg et al. 2006). To help resolve this concern, the US and Mexico have jointly invested in an international treatment plant that removes and treats dry weather flow from the Tijuana River before it drains to the ocean on the US side of the border. Local governments on the US side have also invested in a high frequency radar array, partly deployed on the Mexican side of the border, to inform beach managers when the predominant flow to US beaches is from Mexican waters. These flow data are used to help determine when beaches on the US side of the border are closed to recreation (Clifton et al. 2007).

1.6 CRITICAL QUESTIONS FOR MICROBIAL MONITORING

While there have been some successes in managing the Great Lakes, there remain many further opportunities for improvement of cross-border cooperation. The goal of this book is to identify the specific cross-border issues of greatest concern for the water management community. The focus is on microbial issues, though other water quality issues will be addressed. The focal questions are:

- How can we quantify and prioritize water-based microbial threats to human health?
- What are the biggest shortcomings in our data collection systems that prevent this prioritization?

This book seeks to explore these critical questions as well as suggest a plan of action for transboundary microbial monitoring. In Chapter 2, Erin Dreelin and Joan Rose provide a summary of the major management frameworks and water quality policies in the Great Lakes basin. Chapter 3 provides an historical perspective of microbial water quality, specifically a comprehensive 1914 IJC bacteriological project that can serve as a model for current and future monitoring studies of a large scale water basin. Erin Dreelin describes the historic study and its implications and relevance to current monitoring efforts in the basin. Chapter 4 explores the affect of societal changes on infectious disease. Syed Sattar and Jason Tetro discuss the changes in society's relationship with water and the resultant effects on waterborne pathogens and disease outbreaks. In Chapter 5, Thomas Malone and David Rockwell examine the Global Ocean Observing System as an example of a monitoring system designed to collect and distribute data to groups that manage, study, or depend on ocean systems. Development of an observing system for the Great Lakes is currently underway (NOAA). Chapter 6 moves on to a management framework and how it can be applied to the Great Lakes region for monitoring pathogens. Dan Deere and Annette Davison describe the hazard analysis and critical control point system (HACCP) approach to management and how it could be applied to the Great Lakes basin. Finally, Chapter 7 presents a plan of action based on the information the authors discuss in each chapter. Joan Rose provides a comprehensive framework for implementing a monitoring and risk assessment strategy in the Great Lakes basin.

1.7 REFERENCES

Blanch, A.R., Belanche-Munoz, L., Bonjoch, X., Ebdon, J., Gantzer, C., Lucena, F., Ottoson, J., Kourtis, C., Iverson, A., Kuhn, I, Moce, L., Muniesa, M., Schwartzbrod, J., Skraber, S., Papageorgiou, G.T., Taylor, H., Wallis, J., Jofre, J. 2006. Integrated analysis of established and novel microbial and chemical methods for microbial source tracking. Applied and Environmental Microbiology 72:5915-5926.

Boehm, AB; Grant, SB; Kim, JH; Mowbray, SL; McGee, CD; Clark, CD; Foley, DM; Wellman, DE. 2002. Decadal and shorter period variability of surf zone water quality at Huntington Beach, California. Environmental Science and Technology 36: 3885-3892.

CDC (US Centers for Disease Control). 2007. Parasitic diseases: about parasites. http://0-www.cdc.gov.mill1.sjlibrary.org/Ncidod/dpd/aboutparasites.htm

Clifton, C., S.Y. Kim and E. Terrill. Using real time observing data for public health protection in ocean waters. Environmental Research In review.

Crump, J.A., Luby, S.P. and Mintz, E.D. 2004. The Global Burden of Typhoid Fever. *Bulletin of the World Health Organization*, Vol. 82, No. 5, pp. 346-353.

EPA GLNPO (Environmental Protection Agency Great Lakes National Program Office). 2005. Great Lakes. http://epa.gov/greatlakes/

Fong, T. Mansfield, L. S., Wilson, D.L. Schwab, D.J., Molloy, S.L. and Rose, J.B. 2007. Massive Microbiological Groundwater Contamination Associated with a Waterborne Outbreak in Lake Erie South Bass Island, Ohio, *Environ. Hlth Perspec.* 115 (6) 1-9.

GAO (Government Accountability Office). 2003. Great Lakes: An Overall Strategy and Indicators for Measuring Progress Are Needed to Better Achieve Restoration Goals. GAO-03-515. http://www.gao.gov/new.items/d03515.pdf.

Gersberg, R.M., M.A. Rose, R. Robles-Sikisaka and A.K. Dhar. 2006. Quantitative detection of Hepatitis A Virus and enteroviruses near the United States-Mexico border and correlation with levels of fecal indicator bacteria. Applied and Environmental Microbiology 72:7438-7444.

Gleick, P.H. 1993. *Water in Crisis A Guide to the World's Fresh Water Resources.* Oxford University Press, NY, NY

Gleick, P.H. 2006. *The World's Water.* Island Press, Washington, DC

Great lakes Information Network (GLIN). 2006a. The Great lakes. http://www.great-lakes.net/lakes/

Great Lakes Information Network (GLIN). 2006b. Satellite Image of the Great Lakes Basin. http://gis.glin.net/maps/

Hildebrand, L.P, V. Pebbles, and D. A. Fraser. 2002. Cooperative ecosystem management across the Canada–US border: approaches and experiences of transboundary programs in the Gulf of Maine, Great Lakes and Georgia Basin/Puget Sound. Ocean and Coastal Management 45: 421-457.

Hou, D; Rabinovici, SJ; Boehm, AB. 2006. Enterococci predictions from a partial least squares regression model can improve the efficacy of beach management advisories. Environmental Science and Technology 40: 1737-1743.

Hrudey, S.E., P. Payment, P.M. Huck, R.W. Gillham, and E. J. Hrudey. 2003. A fatal waterborne disease epidemic in Walkerton, Ontario: comparison with other waterborne outbreaks in the developed world. *Wat. Sci. Tech.* 47 (3): 7-14.

Mac Kenzie WR, Hoxie NJ, Proctor ME, Gradus MS, Blair KA, Peterson DE, Kazmierczak JJ, Addiss DG, Fox KR, Rose JB, and David JP. 1994. "A Massive Outbreak in Milwaukee of *Cryptosporidium* Infection Transmitted Through the Public Water Supply." *New Engl. J. Med.* 331(3): 161-167.

National Research Council. 2004. Indicators for waterborne pathogens. National Academies Press. Washington, DC.

NRDC (Natural Resources Defense Council). 2007. Testing the waters 2007: A guide to water quality at vacation beaches. http://www.nrdc.org/water/oceans/ttw/titinx.asp

Nevers, MB; Whitman, RL. 2005. Nowcast modeling of Escherichia coli concentrations at multiple urban beaches of southern Lake Michigan. Water Research 39:5250-5260.

Noble, R.T. and S.B. Weisberg. 2005. A review of technologies being developed for rapid detection of bacteria in recreational waters. Journal of Water and Health 3:381-392.

Noble, R.T., S.B. Weisberg, M.K. Leecaster, C.D. McGee, J.H. Dorsey, P. Vainik and V. Orozco-Borbón. 2003. Storm effects on regional beach water quality along the southern California shoreline. Journal of Water and Health 1:23-31.

Ocean.us. 2007. Public health risks: Coastal observations for decision-making. Publication #15. The National Office for Integrated and Sustained Ocean Observations. Washington DC.

Rabinovici, S. J. M., R. L. Bernknopf, A. M. Wein, D. L. Coursey, and R. L. Whitman. 2004. Economic and Health Risk Trade-Offs of Swim Closures at a Lake Michigan Beach, Environmental Science and Technology 38 (10), 2737-2745.

Shiklomanov, I.A. - 2000 - World freshwater resources UNESCO, International Hydrological Programme

Shmueli, D.F.. 1999. Water quality in international river basins. *Political Geography* 18: 437-476.

Wade, T.J., R.L. Calderon, E. Sams, M. Beach, K.P. Brenner, A.H. Williams and A.P. Dufour. 2006. Rapidly measured indicators of recreational water quality are predictive of swimming-associated gastrointestinal illness. Environmental Health Perspectives 114:24-28.

Wolf, A. T., Natharium J.A., Danielson, J.J., Ward, B.S., and Pender, J.K. 1999. International River Basins of the World, Intern. J. of Water Res Develop. 15 (4) http://www.transboundarywaters.orst.edu/publications/register/index.html#toc

World Health Organization 2007. The World Health Report 2007: A Safer Future: Global Public Health Security in the 21st Century

2

Transboundary Water Quality Management in the Great Lakes

E. A. Dreelin and J. B. Rose

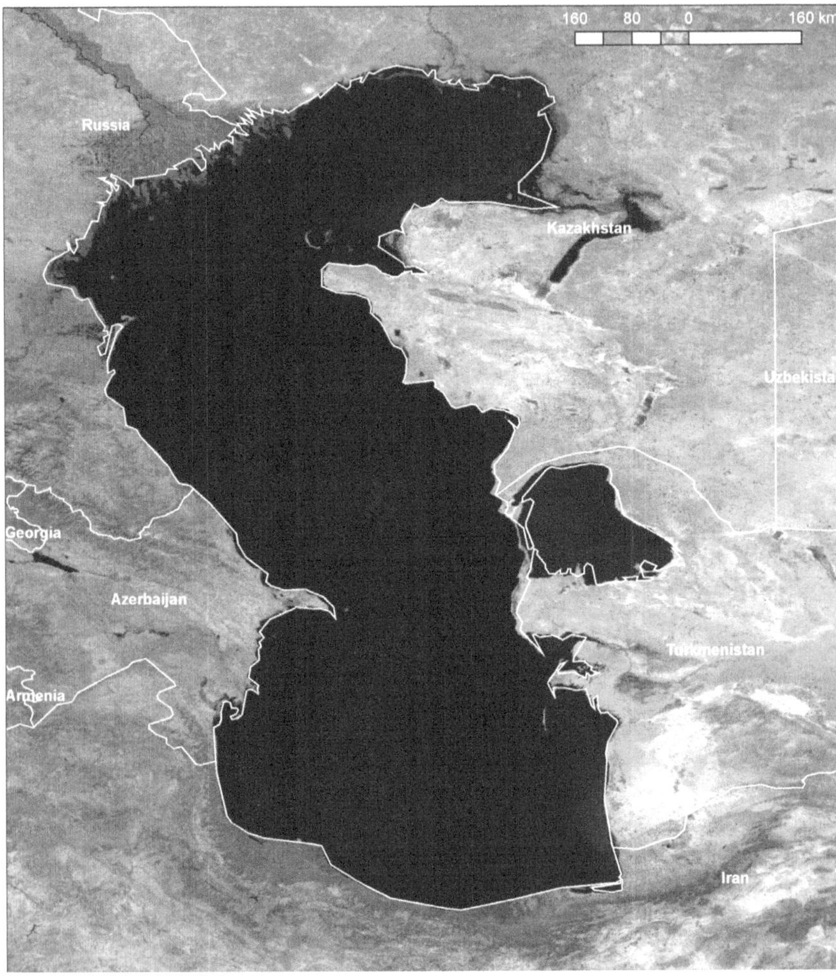

The Caspian Sea. Image: MSU Center for Global Change and Earth Observations

"With so much that is controversial in the field of international relations, it is good to know that two countries with an extended common boundary can live as good neighbours, prepared to consider each other's viewpoints in seeking to resolve problems of mutual interest."

- John Diefenbaker, Prime Minister of Canada, 1961

2.1 INTRODUCTION

Humans are forced to share water resources. Because watershed boundaries rarely align with political borders, effective management of transboundary waters is both contentious and elusive. Conflicts over shared waters occur at all spatial scales, from villages to countries. Despite such difficulties, the Great Lakes basin enjoys a storied history of international cooperation. Although new conflicts over water quality and quantity continue to arise, the Great Lakes region benefits from a legacy of consensus regarding the need for cooperation that facilitates the resolution of emerging problems.

In this chapter, we summarize the major policies and legislation that play a role in managing transboundary water resources between Canada and the United States of America (US) in the Great Lakes basin. Two federal governments, eight US states, two Canadian provinces, several bi-national agencies, native peoples, and many local governments are involved in managing water resources in the Great Lakes. There are numerous treaties and agreements in play in the basin, such as those concerning fisheries management. Here, we will briefly describe only the foundational agreements and subsequent actions that explicitly address water quality management. For more in depth discussions of these agreements, we refer readers to several resources (Table 2.1).

Table 2.1. Resources for further exploration of Great Lakes treaties, legislation, and agreements.

Topic	Source	Comments
Boundary Waters Treaty and Great Lakes Water Quality Agreement	Botts, L. and P. Muldoon. 2005. Evolution of the Great Lakes Water Quality Agreement. Michigan State University Press, East Lansing, MI.	Botts and Muldoon provide a thorough examination of the GLWQA from its creation to subsequent amendments. The authors also discuss the Boundary Waters Treaty as the foundation of the GLWQA and the changing role of the IJC.
International Joint Commission	International Joint Commission website. http://www.ijc.org/	The IJC website provides an overview of the organization, links to the full text of binational treaties and agreements, and online access to IJC publications. Products are available in English or French.
US Clean Water Act	EPA. Introduction to the Clean Water Act. http://www.epa.gov/watertrain/cwa/	EPA's Watershed Academy provides an online overview of the Clean Water Act.
BEACH Act	EPA. 2006. Implementing the BEACH Act of 2000: A report to Congress. http://www.epa.gov/waterscience/beaches/report/	This report to US Congress by EPA summarizes the current status of implementing BEACH Act requirements.
Diversions and consumptive uses	Annin, P. 2006. The Great Lakes Water Wars. Island Press, Washington D.C.	This book discusses the issues involving use of Great Lakes water within and outside the basin, the Great Lakes compact, and legislation in Ontario and Quebec.
Great Lakes Regional Collaboration	Great Lakes Regional Collaboration http://www.glrc.us/	Web site provides information on the recent collaborative effort in the US to protect and restore the Great Lakes

2.2 BOUNDARY WATERS TREATY

The Boundary Waters Treaty, signed in 1909, was a landmark treaty for the US and Canada. The treaty established a means to address cross-boundary water issues and resolve disputes between the two nations. The boundary waters covered under the treaty run the entire length of the US-Canada border, incorporating more than the Great Lakes. Boundary waters are defined as "waters from the main shore to main shore of the lakes and rivers and connecting waterways, or portions thereof, along which the international boundary between the United States and the Dominion of Canada passes, including all bays, arms, and inlets thereof, but not including tributary waters..." (Boundary Waters Treaty 1909). The treaty addresses navigation, water use, and water quality of these boundary waters.

In Article I of the treaty, both US and Canada agree to maintain forever the navigability of all currently navigable boundary waters forever for the benefit of both countries. The article applies to inhabitants and vessels of the US and Canada equally. This agreement also extends to Lake Michigan and the canals connecting boundary waters, even though they do not meet the definition of a boundary water. Each country may adopt rules and regulations, such as charging tolls, to operate canals but those rules and regulations must apply equally to the citizens and vessels of US and Canada. Neither country may discriminate against the citizens or vessels of the other.

Article II sets the framework for dealing with water diversions and reasserts equal rights and privileges for the US and Canada. Jurisdiction over diversions and use of all waters on their respective side of the border remains with the two countries or the states and provinces. Thus, the treaty recognizes that each country, or state/province, has primacy over its own water resources. However, diversion or use of water can not result in any injury on the other side of the border. If injuries do arise, the injured party has the same legal rights and remedies as if they were citizens of the country causing harm. Again, citizens of the US and Canada are treated equally.

The establishment of the International Joint Commission (IJC) is a centrepiece of the Boundary Waters Treaty. The treaty created the IJC to resolve disputes between the countries, review and approve diversion and obstructions, and examine referrals from the two countries. In Article III, future diversions and obstructions on boundary waters require approval of the IJC. The restriction on diversions and obstructions does not cover actions to maintain navigability such as dredging channels, improving harbours, and other government activities that improve commerce. The IJC also reviews and approves projects that would restrict flow of water to the other side of the boundary.

The IJC comprises six commissioners, three from Canada and three from the US. The commissioners are appointed by their respective governments. Decisions are based on majority-rule; if commissioners disagree, the dissenting commissioners may prepare separate reports to the countries. If there is an even split between the commissioners, the two countries are required to revise the question given to the IJC. The treaty sets out guiding principles on which the

IJC bases its decisions. These guiding principles are (Boundary Waters Treaty, Art. III, 1909):

- The two countries have equal rights in the use of boundary waters
- The IJC will use the following order of preference among uses:
  - Domestic and sanitary uses
  - Navigation
  - Power and irrigation

The IJC can also require either country to take actions to mitigate any injury caused by a diversion or use to protect against injury on the other side of the border.

US and Canada can refer matters of dispute to the IJC. The IJC is then required to study the dispute and report its findings to the two countries in writing. The reports of the IJC are not final decisions on the referral; ultimate action is up to the two countries. If the IJC cannot resolve a dispute, the matter can be referred to an umpire under the Hague Convention. In examining referrals, the IJC has the authority to administer oaths to witnesses, issue subpoenas, and take evidence. For matters in which the countries reach an agreement, the treaty requires reciprocal legislation supporting the agreement to be passed by each country.

Although water use and diversion is a main focus of the treaty, water quality is addressed. The treaty prohibits polluting boundary waters, specifically pollution that would cross the boundary and affect the health or property of the other side. As with water use and diversions, any disputes over cross-boundary pollution can be referred to the IJC. Thus, the Boundary Waters Treaty laid the foundation for relations between Canada and the US concerning water resources.

2.3 GREAT LAKES WATER QUALITY AGREEMENT

Canada and the US signed the first Great Lakes Water Quality Agreement (GLWQA) in 1972, with renewal in 1978 and further amendment in 1987. All agreements stress consensus on the shared importance of the Great Lakes to both nations and commit to protecting and restoring the chemical, physical, and biological integrity of the ecosystem. The concluding remarks in the preamble summarize the intent of the agreement:

> *"...the best means to preserve the aquatic ecosystem and achieve improved water quality throughout the Great Lakes System is by adopting common objectives, developing and implementing cooperative programs and other measures, and assigning special responsibilities and functions to the International Joint Commission."*

The 15 articles and 17 annexes of the Agreement describe the goals and objectives for all Great Lake waters. The three major goals are: 1) to prohibit and virtually eliminate the discharge of toxic substances, 2) to provide public assistance to construct publicly owned treatment works, and 3) to coordinate planning by Great Lakes jurisdictions to address all sources of pollutants (GLWQA, Art. II). More specifically, the agreement states that waters should be free of substances resulting from human activity that settle and harm wildlife, that cause color, odor, taste and other conditions that affect beneficial uses, that are toxic to humans or wildlife, and free from nutrients that cause algal blooms (GLWQA, Art. III).

The Agreement establishes specific objectives as minimum measures for both countries (GLWQA, Art. IV). The US and Canada must develop regulations that will, at the very least, meet these objectives. In addition, both countries are required to prevent further degradation of the boundary waters. The Agreement states that dilution is not the solution to pollution. Consequently, increasing flow is not acceptable as a means to lower pollutant concentrations or meet water quality standards. US and Canada agreed to develop programs with the states and provinces to meet the objectives of the agreement. These programs seek to control pollution from municipal sources, industrial sources, agriculture, forestry, dredging activities, shipping, and onshore and offshore facilities. The agreement also calls for programs creating an inventory of pollution abatement requirements, reduction of nutrients, development of contingency plans, virtual elimination of toxic substances, and monitoring. The Agreement also includes development of Remedial Action Plans (RAPs) for Areas of Concern, Lakewide Management Plans (LaMPs) to address critical pollutants, and measures to address pollution from contaminated sediments and groundwater. Details for each of the programs are given in several Annexes to the Agreement.

The IJC was given responsibilities for implementing the agreement (GLWQA, Art. VII). These include:

- Collecting, analyzing and distributing data supplied by the states, provinces, and federal governments regarding water quality of the boundary waters;
- Collecting, analyzing, and distributing data and information on meeting the goals and objectives of the agreement;
- Providing advice and recommendations on water quality problems, regulatory programs, and intergovernmental agreements;
- Providing assistance in coordinating activities under the Agreement;
- Providing assistance and advice on research activities in the Great Lakes; and

- Investigating matters referred to the IJC by the federal governments.

The IJC is required to produce biennial reports on progress made in achieving the goals and objectives of the Agreement and its Annexes. The IJC assesses compliance with the Agreement and progress, or lack thereof, made on all the objectives. In its activities under the Agreement, the IJC can exercise any of the powers granted to it under the Boundary Waters Treaty and subsequent federal legislation. The Agreement also gives the IJC the authority to verify data provided to it by the states, provinces, and federal governments. All information must be shared among the parties.

In order to assist the IJC in carrying out its duties under the Agreement, two boards were formed: the Great Lakes Water Quality Board and the Great Lakes Scientific Advisory Board (GLWQA, Art. VIII). Members of both boards are appointed by the IJC following consultation with the two countries. The boards must contain equal representation from US and Canada. Other Articles of the Agreement describe information sharing and exchange, consultation and review, implementation, rights and obligations of the two countries, the amendment process, when the Agreement goes into effect, termination of the Agreement, and suppression of earlier Agreements.

Major activities described in the Annexes include the virtual elimination of pollutants, Remedial Action Plans to address the most contaminated sites in the basin, Lakewide Management Plans, and phosphorus management. Annexes also include programs to address pollutants from shipping vessels, dredging, hazardous pollutants, persistent toxic substances, nonpoint source pollution, contaminated sediment, contaminated groundwater, and research and monitoring. The first Annex gives specific limits for chemical, physical, microbiological, and radiological characteristics of Great Lakes waters. Chemical objectives include specific concentration goals for persistent toxins, inorganic compounds, non-persistent toxins such as organic pesticides, and other substances. The physical objectives address asbestos, temperature alterations, and suspended solids. The microbiological objective states that "waters used for body contact recreation should be substantially free from bacteria, fungi, or viruses that may produce enteric disorders or eye, ear, nose, throat, and skin infections or other human diseases and infections."

The second Annex requires development of Remedial Action Plans (RAPs) and Lakewide Management Plans (LaMPs). Remedial Action Plans are required for all Areas of Concern. An Area of Concern is a site that fails to meet the General or specific objectives such that beneficial uses of the water are impaired. There are currently 43 Areas of Concern, two of which have been de-listed (Fig 2.1).

Figure 2.1. Map of the Areas of Concern in the Great Lakes basin. Source: Environment Canada, http://www.on.ec.gc.ca/water/raps/map_e.html.

Areas of Concern are designated by the federal governments with input from the states, provinces, and IJC. The IJC reviews and make recommendations regarding progress addressing the problems in the Areas of Concern. Remedial Action Plans require the federal government(s) to work with state/provincial and local governments to develop and implement the plan. RAPs detail the type of environmental problem, identify the beneficial use(s) impaired, identify causes of impairment, evaluate measures in place to address the impairment, examine alternative remediation measures, discuss selected remediation measures, propose a schedule for implementation, identify responsible parties for each part of implementation, evaluate procedures taken for remediation, and track progress towards completion.

The LaMPs address critical pollutants in each of the Great Lakes. The plans include information similar to RAPs and require governments to:

- Identify threats to human health and wildlife
- Evaluate available information on concentration, sources, fate, and transport of critical pollutants

- Identify steps to reduce pollutant loads and estimate of total loads
- Determine load reductions needed to met specific objectives
- Evaluate management measures for controlling pollutants
- Identify any additional remediation measures necessary to reduce pollutant loads
- Identify responsible parties for each component of implementation
- Evaluate progress
- Develop monitoring program for critical pollutants as well as implementation of plan
- Develop methods for stating when critical pollutants are absent from lake waters.

Both the RAPs and LaMPs use an ecosystem management approach. Reports on the status and progress under RAPs and LaMPs are due to the IJC every two years. The full text of all the RAPs and LaMPs and biennial reports are available online from the US Environmental Protection Agency (EPA) Great Lakes National Program Office and Environment Canada (http://www.epa.gov/glnpo/).

The Water Quality Agreement was recently reviewed in 2006 and a number of key recommendations were identified (IJC 2006). These recommendations included improving accountability, revising the Agreement so that it is action oriented, involving more partners, incorporating an ecosystem approach to management, and defining human health, among others. Because there have been many advances in the scientific data and methods available, it was recommended that goals and objectives be revised and that the annexes be updated to address emerging problems. In addition, the direct connection between water quality and public health needed to be restated and specifically articulated. No where in the current Agreement is the word "pathogen" defined and addressed. The recommended revisions would provide a management framework based on a process that includes a strong commitment to research, monitoring, data management, assessment and communication.

2.4 NATIONAL WATER QUALITY PROGRAMS

2.4.1 United States

The Clean Water Act (CWA) is the major federal legislation regulating ambient water quality and discharges in the United States. Before delving into the CWA, it is important to consider the different roles of the US federal and state governments when it comes to environmental programs. In the US, the Constitution does not explicitly discuss environmental protection. The source of authority for federal environmental programs is the Commerce Clause. The Commerce Clause gives authority to regulate interstate commerce to the US

Congress. The intent is to ensure a level playing field amongst the states and prevent states from discriminating against one another. States have police powers, namely the power to protect public health, safety, and welfare. Also, any powers not expressly assigned to the federal government by the Constitution remains with the states or people.

In the 1960s and 1970s, awareness of environmental problems increased as Americans witnessed events such as rivers catching on fire and books such as Rachel Carson's Silent Spring caught national attention. A groundswell of popular support for environmental programs spurred the federal government to take action. Prior to the signing of the Clean Water Act, federal water quality programs mainly provided funding to states for constructing municipal treatment works. Signed in 1972, the Federal Water Pollution Control Act (Clean Water Act) set out two objectives 1) to eliminate all discharges to navigable waters by 1985 and 2) to achieve interim water quality that was protective of fish, shellfish, wildlife and recreation. The Act has been amended and reauthorized since 1972, with major amendments occurring in 1977 and 1985. The states implement most CWA programs with federal oversight. The Act has many programs, but we are only going to discuss those most relevant to microbiological monitoring, namely the water quality standards and the National Pollutant Discharge Elimination System programs.

The Act requires states to establish water quality standards. Water quality standards have three components: 1) a designated use for the water body, 2) water quality criteria necessary to maintain that use, and 3) an anti-degradation policy. A designated use is defined by the state and describes the activities a waterbody should be able to support, such as full body contact recreation, fishing, or drinking water supply. Criteria are then set for meeting the designated use. Criteria may either be narrative, describing certain conditions to attain, or quantitative, giving specific numerical limits for a pollutant. The criteria are what many people think of when hearing 'water quality standard.' EPA recommends water quality criteria and provides guidance to the states, but it is the states which actually adopt legally-binding criteria for ambient waters. A state can adopt the EPA criteria, a more stringent standard, or develop its own standard. EPA has final approval for accepting the state criteria. The final component of a water quality standard is the anti-degredation policy. Anti-degredation requires states to protect existing uses, prevent water quality from further degrading, and prohibits removing a designated use to weaken water quality standards. The bacterial fecal indicator *E.coli* is used most often as the recreational standard for microbiological water quality in fresh water systems in most of the states and provinces. Thus, *E. coli* is used to assess microbiological quality and to determine impairment of designated uses. The need to identify the source of *E.coli* in impaired waters has been rising in prominence in order to develop management strategies.

The National Pollutant Discharge Elimination System (NPDES) requires that all entities discharging to navigable waters receive a discharge permit (Section 402). NPDES permits set limits on the quantity and quality of the discharge and requires monitoring to show compliance with the permit. All but five states, Washington D.C., and US territories have an authorized NPDES program (EPA 2007). Standards for effluent limits are either technology or water quality-based. The technology-based standards do not specify which technologies must be implemented, rather they state a concentration or specific standard for effluent that must be met. The limit is based on what current technologies can achieve and also consider cost of implementation. These are performance-based standards that limit the amount of pollutants coming out of the end of a pipe. In addition, the Clean Water Act also uses standards based on the water quality of receiving waters. These water-quality based standards focus on the amount of pollution a waterbody can receive and still meet its designated use. Over time, the program has increasingly focused on limiting impacts on waterbodies. Thus, where early efforts by EPA largely dealt with developing technological standards, later efforts have focused on receiving waters and water-quality based standards.

Over time, CWA programs have changed from assessing what comes out of the end of a pipe to assessing what enters the waterbody from a variety of sources, including nonpoint source runoff, and the resulting receiving water quality. NPDES permits for sewage treatment plants that discharge to recreational waters typically include effluent limits for fecal coliform bacterial concentrations, achieved primarily through disinfection. In some states this limit may coincide with the swimming season and in other areas, no disinfection is used for discharges to waters that are used solely for secondary recreational purposes (e.g., boating, fishing, non-contact activities).

In 2000, the US Congress passed the Beaches Environmental Assessment and Coastal Health Act (BEACH Act) as an amendment to the CWA. The BEACH Act was a response to reports of declining water quality at recreational beaches. The Act applies only to coastal recreational waters as defined by the Act, namely the Great Lakes and marine coastal waters designated as recreational waters by a state. In 1986, EPA recommended a criteria of 126 *E. coli*/100 mL for freshwater based on a 30-day geometric mean using at least 5 samples (EPA 1986). The BEACH Act required coastal states to adopt standards at least as protective as the 1986 EPA criteria by 2004. If the states failed to adopt standards, EPA would set the standards for the state. The BEACH Act also required EPA to conduct more research regarding human health and waterborne pathogens and to develop new criteria for pollution indicators based on the findings. Finally, the Act provided funding to state and local governments to implement beach monitoring programs.

2.4.2 Canada

In Canada, the provinces have primary jurisdiction over protection of water quality. However, under the Canada Water Act, the federal government may become involved in issues of national significance. The Canada Water Act, first passed in 1970, provides a framework for the federal and provincial governments to enter into agreements for management of water resources. These agreements establish procedures for monitoring, data analysis, research, development of detailed management plans, projects for conservation efforts, and implementation of management plans. The agreements include details on who (i.e., federal or provincial government) is responsible for each activity under the agreement, costs, a schedule for activities, and terms and conditions of the agreement. In essence, these agreements enable the federal government and provinces to develop plans and conduct research in areas of national concern.

The Canada Water Act also enables the federal government to establish water quality management areas. The Minister of the Environment may designate specific geographic areas as water quality management areas and name a water management agency for the area based on recommendation by the federal cabinet. Designation as a water quality management area triggers regulations regarding pollution of waters. Discharge of waste to water management areas is prohibited or must be in accordance with the water management plan.

The National Guidelines and Standards Office (NGSO) of Environment Canada develops national water quality guidelines under the Canadian Environmental Protection Act and agreements such as the Great Lakes Water Quality Agreement. These guidelines are similar to the criteria that EPA develops as guidance to the states. Working with partners, the NGSO has developed water quality guidelines for protection of aquatic life, agricultural water uses, drinking water, and recreation (NGSO 2007). The national guidelines for recreational water in freshwater is based on the indicator, *E. coli* (Federal-Provincial Working Group on Recreational Water Quality 1992). The guideline states that the maximum value based on a 30-day geometric mean of at least 5 samples should not exceed 2000 *E. coli* /L (200 *E. coli*/100mL).

2.5 STATE AND PROVINCIAL PROGRAMS

2.5.1 Great Lakes States Compact

Although each state is responsible for its own water quality and quantity programs, the states can enter into agreements with one another. The mechanism for this in the US is through a compact. The US Constitution states that "No state shall, without the consent of Congress...enter into any agreement or compact with another state, or with a foreign power..." An interstate compact must be approved by each state legislature, the governor of each state, and the US Congress.

The Great Lakes states (Illinois, Indiana, Michigan, Minnesota, New York, Ohio, Pennsylvania and Wisconsin) passed the Great Lakes Compact in 1955 and it was given congressional approval in 1968. As stated in Article I of the Compact, the purpose of the compact is:

> *"1. To promote the orderly, integrated, and comprehensive development, use, and conservation of the water resources of the Great Lakes Basin (hereinafter called the Basin). 2. To plan for the welfare and development of the water resources of the Basin as a whole as well as for those portions of the Basin which may have problems of special concern. 3. To make it possible for the states of the Basin and their people to derive the maximum benefit from utilization of public works, in the form of navigational aids or otherwise, which may exist or which may be constructed from time to time. 4. To advise in securing and maintaining a proper balance among industrial, commercial, agricultural, water supply, residential, recreational, and other legitimate uses of the water resources of the Basin. 5. To establish and maintain an intergovernmental agency the end that the purposes of this compact may be accomplished more effectively."*

The compact established the Great Lakes Commission (GLC) as the intergovernmental, regional agency to represent and work with the states. The compact also stated that Ontario and Quebec could join the compact, as their laws allow. In fact, the term 'state' was formally extended to include these provinces in the compact. In 1999, a Declaration of Partnership gave Ontario and Quebec Associate Member status in the compact. Thus, the eight states and two provinces all have representatives on the GLC.

The GLC examines issues related to water resources and makes policy recommendations to the states and provinces. However, these recommendations are not legally-binding. It is up to the states and provinces to act on the recommendations.

2.5.2 Canada-Provincial Agreements

Both Ontario and Quebec have entered into agreements with the federal government of Canada regarding water resources in the Great Lakes Basin. The Canada-Ontario Agreement Respecting the Great Lakes Basin Ecosystem coordinates planning, management, and restoration efforts by the two governments to improve the ecosystem of the Great Lakes. The first Canada-Ontario Agreement was signed in 1971 in response to early agreements between Canada and the US regarding phosphorus (Environment Canada 2007). Subsequent agreements were signed focusing on other aspects of water quality. Currently, Canada and Ontario are working on the draft 2007 Agreement. The current agreement, signed in 2002, established guiding principles, describes management strategies, and provides an administrative framework for implementing the Agreement (Canada-Ontario Agreement 2002).

The Canada-Quebec Agreement was first signed in 1989, and has developed into the St. Lawrence Action Plan. The Agreement, like the Canada-Ontario Agreement, states how Quebec and Canada will cooperate to monitor, research, protect, and restore the St. Lawrence River basin. The plan includes objectives to reduce toxins, protect biodiversity, protect human health, address pollution from different land uses, and involve the community. Canada and Quebec recently developed the St. Lawrence Plan for a Sustainable Development 2005-2010. This plan extends the agreement and work of the two governments to conserve and protect the St. Lawrence River basin.

2.6 PATHOGEN ISSUES IDENTIFIED

As mentioned earlier, the Agreement does not explicitly mention waterborne pathogens. However, a number of key pathogens and issues have arisen in the Great Lakes basin. These issues include:

- Recalcitrant, potent and emerging pathogens such as: *Cryptosporidium, E.coli 0157:H7; Arcobacter, Campylobacter*, antibiotic resistant bacteria, *Helicobacter*, adenoviruses, noroviruses and toxic Cyanobacteria;
- Increases in beach advisories and closures;
- Contamination of sediments and sand and the potential for sediments and sand to serve as reservoirs for pathogens;
- Development of microbial source tracking methods and applications;
- The need for a waterborne disease registry;
- The impact of untreated sewage on the Great Lakes;
- Septic tank impacts on surface water quality;

- Concentrated feeding operations and safe manure application to land;
- The continuing problem of storm water and nonpoint source pollution; and
- Climate change.

Currently, there is no framework to identify the sources of pathogens, types of pathogens, or the risk to human health as a result of exposure via the water environment. The fecal bacterial indicator system is severely limited in its ability to address pathogen risks of the 21st century. In a large international basin, transboundary monitoring that can provide results at a meaningful scale is needed. Data need to be shared and communicated among all partners in the basin as is recognized in the Agreement. And finally, research efforts should be coordinated so that the scientific information can be used in a way that will allow for synergistic opportunities for management of the human health risks.

2.7 REFERENCES

Canada-Ontario Agreement Respecting the Great Lakes Basin Ecosystem. 2002. http://www.ene.gov.on.ca/envision/water/greatlakes/coa/1359COA_annex.pdf

Environment Canada. 2007. The 1994 Canada-Ontario Agreement Respecting the Great Lakes Ecosystem. http://www.on.ec.gc.ca/laws/coa/coa94-e.html.

EPA (United States Environmental Protection Agency). 1986. Ambient Water Quality Criteria for Bacteria – 1986. Office of Water, EPA440/5-84-002.

EPA (United States Environmental Protection Agency). 2007. State program status. http://cfpub.epa.gov/npdes/statestats.cfm?program_id=12

Federal-Provincial Working Group on Recreational Water Quality of the Federal-Provincial Advisory Committee on Environmental and Occupational Health. 1992. Guidelines for Canadian Recreational Water Quality. http://www.hc-sc.gc.ca/ewh-semt/water-eau/recreat/index_e.html

IJC (International Joint Commission). 2006. Advice to governments on their review of the Great Lakes Water Quality Agreement. http://www.ijc.org/rel/pdf/advicefinalwc.pdf

National Guidelines and Standards Office (NGSO). 2007. Canadian Water Quality Guidelines. http://www.ec.gc.ca/ceqg-rcqe/English/ceqg/water/default.cfm.

3

Bacteriological Monitoring in the Great Lakes: A Historical perspective to inform the Present

E. A. Dreelin

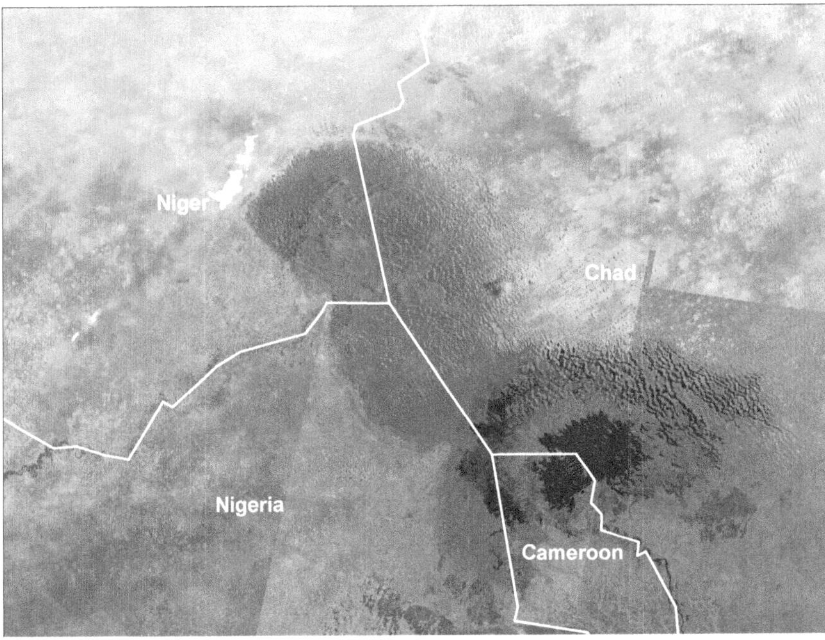

Lake Chad. Image: MSU Center for Global Change and Earth Observations

3.1 INTRODUCTION

In 1909, the United States and Great Britain (on behalf of Canada, which became independent in 1931) signed the Boundary Waters Treaty aimed at preventing and resolving disputes between US and Canada over waters forming the boundary between the two countries and which also established a formal binational body, the International Joint Commission (IJC; See Chapter 2). The treaty was originally intended to protect lake levels and navigability but it has also been the basis for the IJC to get involved in pollution problems in the boundary waters (Binder 1972). Article IV of the treaty states that the "boundary waters and waters flowing across the boundary shall not be polluted on either side to the injury of health or property on the other."

The interest in water quality and impacts on human waterborne disease instigated one of the most comprehensive bacteriological studies ever conducted in the Great Lakes region in 1913 by the IJC at the request of the two governments to address Article IV of the 1909 treaty. This study is possibly the most extensive characterization of bacterial pollution ever conducted, yet few scientists know of it (Durfee and Bagley 1997). Durfee and Bagley (1997) discuss the history of the study and the draft treaty based on the study results that were never ratified; here I provide a review of the methods and major findings, addressing implications for current monitoring programs.

3.2 THE 1914 IJC BACTERIOLOGICAL STUDY

In 1909, Canada and the US charged the IJC with determining the extent of cross-boundary pollution in the Great Lakes (IJC 1914). The two specific questions the IJC were to address were:

(1) "To what extent and by what causes and in what localities have the boundary waters between the US and Canada been polluted so as to be injurious to public health and unfit for domestic and other uses? and,
(2) In what way or manner, whether by construction and operation of suitable drainage canals or plants at convenient points or otherwise, is it possible and advisable to remedy or prevent the pollution of these waters, and by what means or arrangement can the proper construction or operation of remedial or preventative works, or a system or method of rendering these waters sanitary and suitable for domestic and other uses be best secured and maintained in order to insure the adequate protection and development of all interests involved on both sides of the boundary, and to fulfill the obligations undertaken in Article IV of the waterways treaty of January 11, 1909 between the US and Great Britain, in which it is agreed that the waters therein defined as boundary waters and waters flowing across the boundary shall not be polluted on either side to the injury of health or property of the other?"

At the time the study was commissioned, seven million people resided along the boundary waters in US and Canada (IJC 1914). The communities within the basin were using the waterways for sewage disposal, discharging untreated municipal sewage into the basin's lakes and rivers (IJC 1914). A report by sanitary experts contained within the 1914 Progress Report (IJC 1914) shows that waters near all of the population centers were polluted, in some instances the pollution extended from shore to shore across the international boundary. In addition to municipal waste, thousands of vessels were also discharging untreated sewage into the waters of the Great Lakes. The US and Canada had been receiving an increasing number of complaints about water quality in the region (Durfee and Bagley 1997) and deaths due to typhoid were on the rise (IJC 1914). Given the magnitude of the pollution and very real threat to human health, the IJC began planning what later was to be called a "flawless" study of microbial contamination in the Great Lakes (Durfee and Bagley 1997).

The IJC established a committee to oversee the study headed by one US and one Canadian representative of the IJC (IJC 1914). The committee asked Allan McLaughlin of the US Public Health Service and T. A. Starkey of McGill

University to consult on the development of the experimental design (IJC 1914). McLaughlin and Starkey essentially acted as the principal investigators for the study under supervision of the IJC committee.

In designing the study, the committee first called in public officials from around the Great Lakes basin to comment on the proposed monitoring plan. These officials from the States and Provinces provided input on the organization of the monitoring plan developed to address the questions of (1) the extent of cross-boundary pollution and (2) strategies for possible remedies. At a meeting of the IJC in 1912, it was agreed that experts should be called in to comment on the scope and extent of the study. In a conference in December of that year, locations where samples should be collected, the types of analyses, and the general methods of the study were determined. The IJC decided "the investigation shall include a bacteriological examination of the samples taken, including bacterial count and the qualitative and quantitative estimation of *B.coli* according to standard methods, and such chemical examination as may be subsequently deemed necessary."

In addition to the input gained from these public officials, the committee and principal investigators in charge of the study also conducted hearings in Buffalo in June 1913 (IJC 1914). These hearings were announced in the local paper and at local health departments. The purpose of the hearings was to examine possible remedies to pollution in the Niagara River. While all parties at the hearings agreed that discharges of raw sewage into the river should cease, they required more time to consider practical and reasonable alternatives for disposal of waste.

3.2.1 The Approach for the Assessment

The study was limited to those areas where cross-boundary pollution was observed or expected (i.e. where pollution on one side of the US-Canadian border was thought at that time to affect waters on the other (IJC 1914)) rather than covering the entire Great Lakes in order to reduce the scope to a more manageable size. The sites were chosen with input from public officials at the December conference and the June public hearings. The five areas focused on the Niagara River, Detroit River and connecting waterways, St. Mary's River, St. Lawrence River, and portions of the St. John's River (Figure 3.1A and B).

A

B

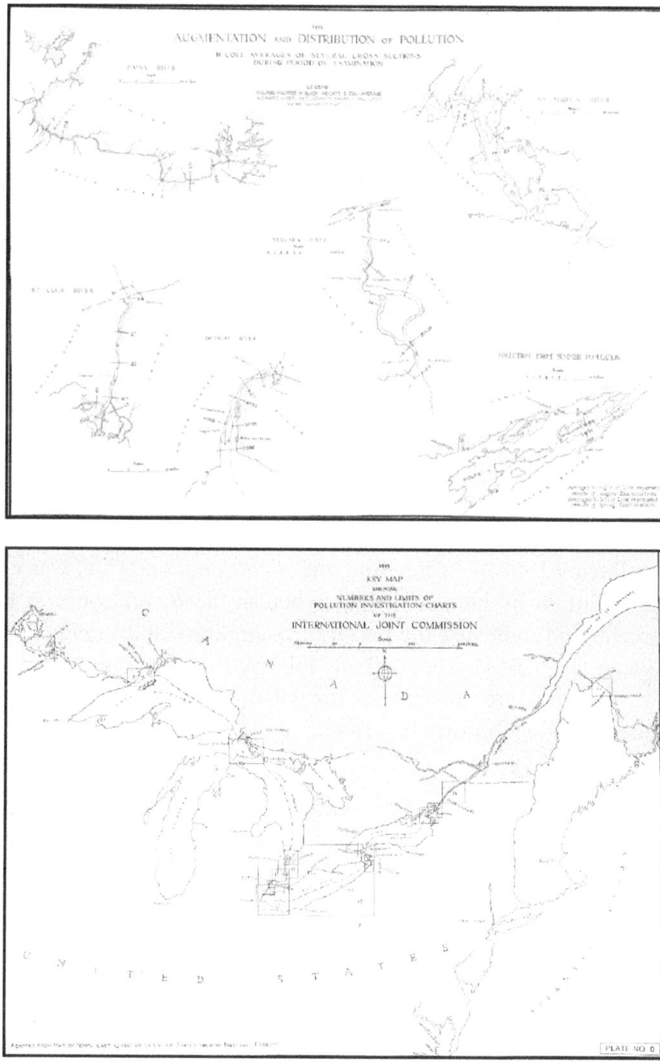

Figure 3.1. (A) Examples of sampling transects for the major rivers samples during the study. Note that transects run perpendicular to the international boundary. Several transects were also located along the boundary in the main shipping channel (IJC 1914). (B) Map of the Great Lakes basin, boxes show areas studied by the IJC (IJC 1914)

The study examined more than 2000 miles of sampling transects and collected over 19,000 water samples. In all, 17 laboratories were involved in the study, each of which was installed, equipped and staffed by the IJC. The US Public Health Service, the Provincial Boards of Health of Ontario and Quebec, the Michigan State Board of Health, and the New York State Department of Health provided assistance to the IJC. Also, many municipalities cooperated by providing information about their water and wastewater treatment practices. The study cost $42,138, which was divided equally between the US and Canada.

A total of 1447 locations were sampled across the Great Lakes during the summer of 1913 (Table 3.1). Most of the samples were collected along transects that ran perpendicular to the US-Canada border or they ran along the boundary, generally in the shipping channels (Figure 3.1). In addition to sampling along these transects, sample at several locations were collected at various depths to characterize how bacterial contamination varied with depth. These depth profiles were conducted in the Rainy, St. Clair, Detroit, St. Lawrence and Niagara Rivers.

For the bacteriological analyses, the investigators determined the total and average number of *Bacillus coli* colonies on nutrient agar at 18-22° C incubated for 48 hours, the total and average number at 37° C incubated for 24 hours, and the estimation of *B.coli* by fermentation in lactose bile incubated for 48 hours at 37° C. The average bacterial count per 100 ml was determined using the Phelps method (Phelps 1907). At the time, the *B.coli* count at 37°C was considered to indicate pollution by human sewage whereas the *B.coli* count at 18-22°C was less specific and indicated the bacterial community and organic matter content of the water (IJC 1914). The methods followed for determining the *B.coli* count at 37°C, which were innovative for 1913, correspond to current sampling techniques for total coliform bacteria.

Table 3.1. Summary of sampling and results from the bacteriological study of boundary waters from the Progress Report (IJC 1914).

Location	Number of sample points	Total number of samples	Time of collection (all 1913)	Classification	Source of Pollution
Rainy River, Rainy Lake Lake of the Woods	192	955	July	3-5	Municipal sewage discharges
Thunder Bay Lake Superior	72	922	July-Aug	3	Municipal sewage discharges
St. Mary's River	133	1065	June- July	5	Municipal sewage discharges, ships
Lake Huron, St. Clair River	187	2326	July-Aug	2	Ships
Lake St. Clair, Detroit River	174	1755	Sept-Oct	1, 5	Ships, Municipal sewage discharges
Lake Erie	114	1306	Sept-Oct	5	Municipal sewage discharges
Upper Niagara River	193	2999	May-July	1	Municipal sewage discharges
Lower Niagara River and western Lake Ontario	101	1178	May-June, Aug	5	Municipal sewage discharges
Eastern Lake Ontario and St. Lawrence River	153	1890	Apr-May, Aug	1	Ships
St. John River between Edmunston and Grand Falls	32	672	Oct	5	Starch factories

In addition to the bacteriological analyses, meteorological data were recorded at the sampling sites. For all municipalities in the study, the area, population, source of water supply, amount of water pumped, and an estimate of sewage discharge were recorded. Each municipality also reported the number of deaths due to typhoid per 100,000 people. For some cities, bacteriological analyses of domestic tap water were conducted following the methods described above for ambient water samples.

Based on the bacterial counts, the investigators created a scale to classify the purity of Great Lakes waters (Table 3.2). A classification of 1 represented relatively pristine waters, with the total *B.coli* count at 37°C less than 2/100ml. Class 2 represented slight pollution (2-10/100 ml); treatment of water for drinking would be required. Class 3 (10-20/100ml) represented significant contamination of water with bacteria, which would also require treatment before drinking. Class 4 (20-50/100 ml) represented serious contamination and in the opinion of the investigators "can not be classified as good raw water (IJC 1914)." The investigators recommended that Class 5 waters (>50/100 ml), those which were considered grossly polluted, could not be used for drinking water, even with treatment. By today's standards, even class 5 waters would not be considered "grossly polluted;" however, the goal in 2007 and 1914 is still the same, namely waters free of bacteriological pollution.

Table 3.2. Classification of Great Lakes waters based on bacterial findings (from IJC 1914).

Classification	Number of *B. coli*/100 ml	Total Bacteria/ml agar (37C)	Notes
1	<2	<10	Relatively pristine
2	2-10	10-25	Slight pollution
3	10-20	25-50	Significant Contamination
4	20-50	50-100	Serious contamination
5	>50	>100	Grossly polluted

3.2.2 Major Findings

At the time of the 1914 study, every municipality in the study area discharged untreated sewage into the rivers and lakes in the basin (IJC 1914). The discharge of untreated sewage was considered the largest source of bacterial contamination to the boundary waters. In some areas, an additional source of pollution was vessel traffic, which also discharged untreated waste into lakes and rivers. High levels of *B.coli* were found in the Detroit River, Niagara River, Rainy River, St. Clair River, St. John River, lower Lake Erie and lower Lake Ontario (IJC 1914). Some seasonality in bacterial counts was observed for a few sites sampled during several months in 1913 (most areas were sampled once during 1913); in general bacterial counts increased from early summer to late autumn (IJC 1914). For example, *B. coli* counts at Lake St. Clair increased in October, despite lower cumulative precipitation, compared to spring (Figure 3.2). The results for the major study areas are summarized below; all results were reported in the Progress Report to the IJC (IJC 1914).

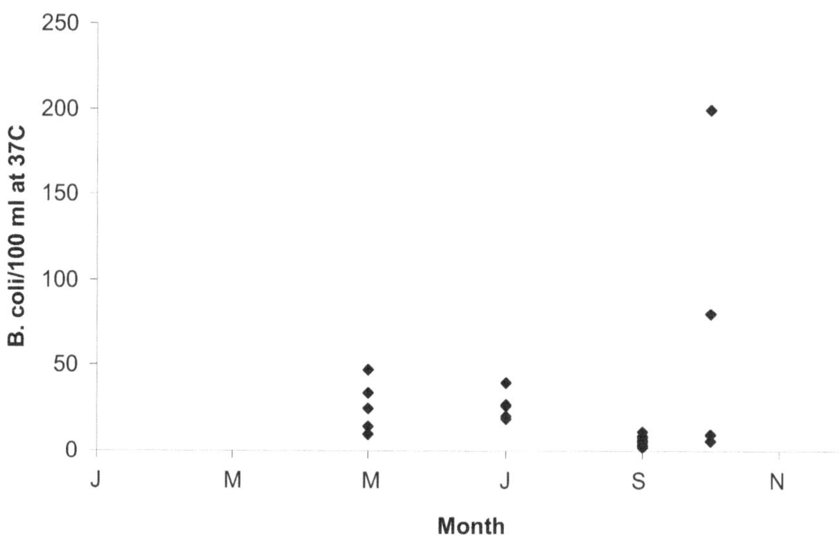

Figure 3.2. Seasonality was observed in *B.coli* results during the 1914 study; data shown are for Lake St. Clair which was sampled during May, July, September, and October of 1913 (data from IJC 1914).

One of the main conclusions of the report was that the municipal water supplies of the Great Lake communities were not safe (IJC 1914). Many municipalities had water intakes located in highly polluted waters and did not have any means to adequately treat the water for drinking. This was reflected in the high rates of typhoid deaths reported across the study area (IJC 1914). The three main reasons given for what the investigators call the "disgraceful" records were:

- Unrestricted discharge of untreated sewage
- Failure to treat or purify polluted water
- Ineffective treatment

In some areas the pollution was so great as to be considered beyond the abilities of the treatment methods in use at the time. Because of this, the IJC (1914) recommended controlling sewage discharge as a means of improving public health.

3.2.2.1 Rainy River, Rainy Lake and Lake of the Woods

Rainy River was expected to be fairly pristine given that the human population in the watershed was low at the time of study (IJC 1914). However, the monitoring results showed that Rainy River, Rainy Lake, and Lake of the Woods were significantly polluted. Upstream waters were classified as Class 3 (10-20 *B.coli* colonies/100ml) and bacterial contamination increased at downstream locations to Class 5 (>50/100ml). One possible explanation for the unexpected contamination was that the Canadian Northern Railway company was working in the area and the worker camps were close to the sampling points (IJC 1914). However, this did not account for the high bacteria counts found throughout the length of the river. Bacterial counts increased with distance down river, particularly near cities, suggesting inputs from sewage. An additional concern was that testing of the domestic water supply in downstream cities showed the same results as the river water, namely that it was contaminated with human sewage.

3.2.2.2 Thunder Bay, Lake Superior

Pollution in Thunder Bay did not cross the boundary and was generally low (Class 2; 2-10 to Class 3; 10-20/100ml) although serious or gross pollution was observed sporadically during the sampling period. Although the study did not find extensive bacterial contamination, local conditions were of concern. Instead of flowing out into the Bay, the investigation found that polluted waters concentrated near shore (IJC 1914). On some sampling dates, the plume of contaminated water ran 9 miles along the shore. The investigators found that the tap water of Port Arthur, which withdrew its water from Thunder Bay, was contaminated.

3.2.2.3 St. Mary's River

The St. Mary's River watershed was sparsely populated at the time of study; however, boat traffic along the river was heavy. Samples taken from the shipping channel had higher counts than those taken between the channel and the shore, suggesting that discharges from vessels were the main source of bacteria. Also, samples taken along a transect running in the channel from Detour Passage to Mackinac Island had coliform present in every sample. The general trend was an increase in *B.coli* counts at sampling locations downstream of towns; in these downstream areas waters were highly polluted (Class 5, >50/100ml). Because of this pollution and the location of the water intakes in areas of high bacteria counts, the researchers recommended treatment of the water before public distribution.

3.2.2.4 Lower End of Lake Huron

The lower end of Lake Huron showed only low levels of pollution, with the majority of samples falling below 10 *B. coli*/100ml (Class 2). However, average bacterial counts increased with distance downstream and close to shore. Nearshore samples on both sides of the bound were typically 15 *B. coli*/100ml (Class 3). Vessels were considered the main source of sewage based on the observation of increased bacterial counts in areas of high boat traffic.

3.2.2.5 St. Clair River

The St. Clair River showed high levels of bacterial contamination due to sewage discharges from the cities Port Huron on the US side and Sarnia on the Canadian side. High bacterial counts were found along the entire transboundary cross-section of the river and the water was classified as grossly polluted (Class 5, >50/100ml). Pollution was worse on the US side, due to large discharges of sewage from Port Huron, than on the Canadian side of the river. The investigators stated that untreated water from the St. Clair River was not safe for domestic use.

3.2.2.6 Lake St. Clair

Lake St. Clair was found to have relatively clean water; mid-lake bacterial counts were generally 6 *B.coli*/100ml (Class 2). However, as seen in other areas examined, counts in the shipping channels and closer to shore were elevated, with measurements occasionally exceeding 100 *B. coli*/100ml.

3.2.2.7 Detroit River

Bacterial counts in the Detroit River were some of the highest observed during the study and the river consistently had *B.coli* counts greater than 50/100ml (Class 5). These waters were considered unsafe as a source of municipal drinking water. The main source of contamination was the City of Detroit. The pollution did cross the boundary and was of considerable concern because water intakes for the Canadian cities of Walkerville and Windsor were downstream of the Detroit sewage discharge. At the time, Detroit used a lime additive to purify its water, which the investigators considered an unreliable technique as typhoid death rates were high even with this treatment measure. In addition, the researchers found that the levels of chlorine required to disinfect the water for the Canadian cities imparted a bad taste to the drinking water.

3.2.2.8 Lake Erie

The western end of Lake Erie was found to be highly polluted (Class 5, >50/100ml). In fact the report states that, "In our opinion, there is no point from the lower end of Lake Huron to the islands which separate the western end from the remainder of Lake Erie from which a safe supply of water could be taken from any considerable portion of the 365 days of the year." In contrast, the central and eastern portions of Lake Erie were found to have low levels of bacterial pollution (Class 1-2, <10/100ml).

3.2.2.9 Upper Niagara River

Upstream of Buffalo, New York, the Niagara River had low levels of *B.coli* (Class 1, <2/100 ml). Levels of bacterial contamination increase downstream of Buffalo and high bacterial counts were observed close to shore. Below Strawberry Island, samples from the entire transect across the river had high bacterial counts (Class 5, >50/100 ml). *B.coli* counts were higher on the US side, presumably due to the Buffalo discharge. Pollution was more uniformly distributed below the falls, demonstrating that the falls mixed rather than removed pollutants as was commonly believed at the time (IJC 1914).

3.2.2.10 Lower Niagara River and Western Lake Ontario

The lower Niagara River was uniformly polluted (Class 5, >50/100 ml), which was likely due to the mixing action of the falls. Very little dilution was found when the contaminated river water entered Lake Ontario and the plume of contaminated water was found as far as 10 miles off shore. One concern about the lack of dilution and mixing once the waters entered Lake Ontario was that passenger vessels often refilled their water supplies in this region. The tap water of several vessels tested showed dangerous levels of bacteria.

3.2.2.11 Eastern Lake Ontario and St. Lawrence River

Eastern Lake Ontario was found to be relatively pristine although the sampling regime showed the negative impacts of vessel traffic. In early samples, taken before the summer boating season, bacterial counts were extremely low (Class 1, <2/100 ml). Samples later in the season showed higher bacterial counts although levels were still relatively low (Class 3, 10-20/100 ml). Water in the

St. Lawrence River was of intermittent quality, at times (later in season) considered grossly polluted.

3.2.2.12 St. John River between Edmunston and Grand Falls

The St. John river was highly polluted (Class 5, >50/100 ml) despite the lack of major cities in the river basin. The elevated bacterial counts found in the St John River were attributed to potato starch factories which discharged waste into tributaries of the river. Human sewage inputs were low due to the small human population size and lack of sewers.

3.2.3 Impact of Study

In order to address strategies to remedy the pollution problem and address the second question charged to them, the IJC convened a panel of experts to advise them on strategies to reduce pollution. The advisory board, composed of public health officials and sanitation engineers, recommended that discharges of untreated sewage into the boundary waters should be prohibited. Interestingly, in areas where only slight levels of contamination (Classes 1-3) were observed, pollution was viewed as inevitable. The advisory board did not recommend sewage treatment in areas where there was no current threat to human health. The board also recommended a discharge limit for *B.coli* as an annual average of 500/100ml. However, given the status of sewage treatment technologies of the time, this discharge limit was considered tentative at best.

Based on the bacteriological study and the advice from the advisory board, the IJC (1918) recommended that "all sewage should, before being discharged into boundary waters, receive some purification treatment, and the degree of such treatment is to be determined in a large measure by the limits of safe loading of a water-purification plant." The IJC also recommended that the requirement for treatment balance concerns of public health and economics.

The Detroit and Niagara Rivers were treated as a separate class, given the high degree of contamination that was found. For the communities discharging sewage into these waters, the IJC recommended that the mean annual cross-sectional average of *B.coli* of the receiving waters not exceed 500/100ml. However, this again was viewed as tentative because of the limits of the current technologies. Because of the limitations in treatment and requirement for bacteriological sampling, a general 'working rule' was offered relying on dilution rather than monitoring bacteria. The rule, based on a series of calculations by one of the engineers on the advisory board, stated that if sewage was diluted in a stream flow of 4 cubic feet/second per capita then the bacterial standard should be reached. The IJC also recommended sewage treatment by fine screening or sedimentation followed by chemical disinfection with calcium hypochlorite or bleaching powder. The 1918 report estimates costs of providing treatment works at $6 million for Detroit, half of which was for construction and the remaining was provided by a 54 cent sewerage charge per capita. This 54 cent charge was viewed as a reasonable cost for the public to bear in order to have safe, clean drinking water (IJC 1918). The final recommendations of the IJC (1918) were to:

(1) Install suitable collecting and treatment works, the latter having special reference to the removal of bacteria and matters in suspension.
(2) Disinfection of vessel sewage before it is discharged.
(3) Develop suitable rules and regulations for protecting the water intakes and ship ballast.
(4) Prohibit discharge of garbage and sawmill waste into boundary waters, and restrict discharge of industrial and other wastes, which are causing appreciable injury, subject to such restrictions as may be prescribed.

It is interesting to note that despite the thoroughness of the bacteriological monitoring, the study was 'lost' to science and even policy. Durfee and Bagley (1997) conducted scientific literature searches for references to the study and only a handful of references were found. Of these references, only one provided a complete citation; the 1914 study suffered from a failure to be cited by later work. Also, the study was never published in the scientific literature and so it remained solely with the IJC.

The IJC prepared a draft treaty at the request of US and Canada based on the findings of the Progress Report (IJC 1914) and recommendations of the Final Report (IJC 1918). While the treaty would have given the IJC authority to regulate pollution of the boundary waters (Bilder 1972), the US and Canada met to negotiate terms but they never reached consensus and the negotiations stopped in 1929; Canada tried to reopen negotiations in 1942 but the US State Department did not think the timing was right given world events (Bilder 1972). Durfee and Bagley (1997) suggest that the failure to implement the findings of the report was due to three major problems:

(1) Failure to communicate the scientific results to the scientific community
(2) The relative obscurity of the IJC at the time
(3) Failure to connect with the most effective level of government, which for the policies being proposed would have been at the state/provincial level rather than federal level.

Given the lack of communication and failure to implement the draft treaty, the 1914 study drifted into obscurity. Fortunately for current and future scientists, the 1914 Progress Report and 1918 Final Report are now available online at the IJC website (www.ijc.org).

3.3 CURRENT MONITORING EFFORTS

The Great Lakes region has not seen another study as comprehensive as the 1914 IJC assessment (Durfee and Bagley 1997). Various agencies at all levels of government currently monitor bacteria in the basin. However, a complete investigation into the extent of bacterial contamination in the Great Lakes has not been conducted recently. Current monitoring efforts include monitoring of drinking water and wastewater treatment facilities and recreational beaches.

3.3.1 Drinking Water and Wastewater

In the US, EPA has set both health goals and legal limits for total coliform levels in drinking water (EPA 2001). The total coliform rule also provides details on the required monitoring protocol for water treatment plants, including the frequency and number of samples to collect. EPA has set the health goal for total coliforms in drinking water at zero. Also, water systems finding coliforms in more than five percent of the samples collected each month fail to meet the total coliform standard. If more than five percent of the samples contain coliforms, water system operators must report the violation to the state. Positive fecal coliform tests may indicate that the system's treatment technologies are not performing properly. Systems may need to take a number of actions to avoid or eliminate contamination when coliforms are found, including repairing the disinfection/filtration equipment, flushing or upgrading the distribution system, and enacting source water protection programs to prevent contamination (EPA 2001). The only limit on coliform bacteria in the ambient water may be found in the exception to the *Surface Water Treatment Rule* (EPA 1989) to avoid filtration of surface waters which states that total coliform bacteria should be less than 100 CFU/100ml or fecal coliform should be less than 20 CFU/100 ml in 90% of the samples collected. These waters would be Class 3 waters following the 1914 IJC classification (Table 3.3); Class 3 waters were considered at the time to have significant pollution and were not recommended as source water for water treatment plants at the time (IJC 1914).

Wastewater treatment plants, or publicly owned treatment works (POTW), also monitor fecal coliform. In the US, fecal coliform is considered a conventional pollutant under the Clean Water Act and so it is controlled through use of technology standards. POTWs must use the best available technology to minimize the amount of fecal coliform in effluent discharged to receiving waters. It is interesting to note that many of the current treatment approaches mirror the approach outlined by the IJC in 1918, namely primary (mechanical) and secondary (biological) treatment of sewage (IJC, 1918).

3.3.2 Beach Monitoring

The US BEACH Act of 2000 requires states to adopt bacteria limits as a part of their state water quality standards that are protective of human health (see Chapter 2). EPA (2004) recently published the final rule for coastal and Great Lakes waters which sets federal standards for ambient bacteria levels. EPA (2004) has recommended a criterion of 126 /100ml for *E. coli*. Michigan and Ohio have already adopted standards as protective as this new EPA criterion, the other Great Lake States are in the process of adopting similar standards (EPA 2004). These waters would be Class 5 waters following the 1914 IJC classification (Table 3.3); which were considered grossly polluted (IJC 1914). Beach monitoring programs are commonly conducted by local government health departments. Potentially, these monitoring programs can provide important information regarding trends in near-shore bacterial contamination. However, long-term trends in the bacterial indicators are difficult to identify because of inconsistencies in monitoring methods and incomplete reporting over time.

Table 3.3. Comparison of guidelines from 1914 IJC bacteriological assessment to modern bacteria standards. The 1914 classifications are based on the assumption that 18% of total coliforms are fecal coliforms (EPA 1986) and therefore only provide a rough comparison.

Application	Guideline	1914 Classification (IJC 1914)
Waters used for drinking to avoid filtration	≤100 CFU/100 ml Total coliform	≤5
Michigan Beach standard	126 CFU *E. coli*/100 ml (monthly geometric mean)	5
EPA Recreational standards	126 CFU *E.coli* / 100 ml	5
Environment Canada Beach standard	200 Fecal coliforms/100ml	5
Sewage Discharges	Varies by state, often 200 Fecal coliforms/100ml	5

In Canada, bacterial contamination of public Great Lakes beaches is regularly assessed during the swimming season by public health authorities, who determine the numbers of *Escherichia coli* (*E. coli*) or similar (fecal coliform) bacterial indicators in the waters near public beaches (Edsall and Charlton 1997). Beach closures indicate high levels of bacterial contamination, at least in near-shore waters off the coast of recreational beaches. However, as in the US, sampling and testing procedures have not been standardized (Edsall and Charlton 1997). Unfortunately, given the gaps and inconsistencies in the data set it is not possible to truly assess trends in bacterial contamination at a given beach or across the Great Lakes. Another complicating factor is the use of varying guidelines for what triggers a beach closing. The new EPA rules (EPA 2004) require an *E. coli* standard of at least 126/100 ml but states are allowed to use a more stringent standard. In Canada, Ontario uses a standard of 100/100ml whereas the national standard recommended by Environment Canada is 200/100 ml (Edsall and Charlton 1997).

3.3.3 Other Monitoring Efforts

There are many current water quality monitoring programs in the Great Lakes basin; however, many of them do not include bacterial analyses. The recent USGS National Water Quality Assessment Program conducted a comprehensive study of water quality in Lake Erie and the St. Clair River (Meyers *et al.* 2000) but fecal coliform bacteria were not included in the parameters monitored. University research programs, such as the University of Wisconsin-Milwaukee WATERbase program (UWM 2003), do monitor bacterial levels but the focus is on free-living bacteria and bacterial production. As human illness in the Great Lakes region due to waterbourne contaminants became rare in the 20th century (Health Canada 1995a, 1995b, 1995c; Health and Welfare Canada 1980), management and monitoring efforts have shifted to focus on other environmental problems such as invasive species, eutrophication, and toxic sediments. There also seems to be a shift in focus to using ecosystem-level indicators in the Great Lakes following the start of the State of the Lakes Ecosystem Conference (SOLEC) program which began in 1994. Microbial pathogens are still a major concern in the Great Lakes basin, particularly given the number of combined sewer overflows in the basin (IJC 2004). Also, new methods to detect the actual pathogens rather than the use of indicators such as total coliform or *E. coli* have been recommended by the IJC (IJC 2004).

3.4 CONCLUSION

There are several lessons to be learned from the 1914 IJC study. One of the reasons the study was so advanced for the time is that the IJC consulted scientists, engineers, and public health officials across the basin for input into the study design (an expert panel was used to give serious review of the study methods and the adequacy of the research questions posed to the IJC by Canada and the US). These advisors promoted the use of the most technologically advanced methods; at the time bacteria samples were often grown in gel rather than agar (Durfee and Bagley 1997). Also, interested parties were given a chance to comment on the study at the Buffalo hearings. Stakeholder participation is one of the essential elements for a successful ecosystem management plan (Hartig *et al.* 1998); the design and successful completion of the 1914 IJC study shows how input from experts and stakeholders can help produce high quality research.

Although the study is a model for involving stakeholders in scientific research, it also shows the need for communicating results and continued engagement with the government. One factor Durfee and Bagley (1997) give as a reason that little came of the treaty based on the results of the study was that the IJC moved on to other pressing problems. In fact, these problems (identified in other references sent to the IJC by the US and Canadian governments) were even mentioned in reports on the bacteriological study (Durfee and Bagley 1997).

Hartig *et al.* (1998) identified eight elements that are required for successful implementation of management plans. These are stakeholder involvement, commitment of top leaders, agreement on information needs and interpretation, action planning within a strategic framework, human resource development, results and indicators to measure progress, systematic review and feedback, and stakeholder satisfaction. The 1914 study did actively involve stakeholders and government officials. Given the failure of negotiations between the US and Canada on the draft treaty, it seems top level officials were not committed to implementing the IJC recommendations. The principal investigators moved on to other projects and apparently did not continue to work on the implementation of the findings or published the work in journals (Durfee and Bagley 1997). While data are gathered now at the county and state level, there has been no comprehensive data base, and the data are fragmented in time and space in the Great Lakes basin. Also, parties in the basin have not agreed to consistent monitoring methods. Thus, our current lack of understanding on the extent of bacterial contamination in the Great Lakes largely is due to a lack of consistency in monitoring, assessment of the data and lack of basin-scale planning. Future efforts can learn from these past and present successes and failures which show the importance of stakeholder engagement, long-term commitment, and communication in developing a successful strategy for managing bacterial pollution in the Great Lakes.

3.5 REFERENCES

Bilder, R. B. 1972. Controlling Great Lakes pollution: A study in the United States-Canadian environmental cooperation. Michigan Law Review 70: 469-556.

Durfee, M. and S. T. Bagley. 1997. Bacteriology and diplomacy in the Great Lakes 1912-1920. Paper prepared for the 1997 meeting of the American Society for Environmental History, Baltimore, MD March 6-9, 1997.

Edsall, T. and M. Charlton. 1997. State of the Lakes Report: Nearshore waters of the Great Lakes. ISBN 0-662-26031-7. www.epa.gov/glnpo/solec/96/nearshore/human_health.html#8.1%20%A0%20Infectious%20Organisms%20as%20Health%20Hazards%A0.

Environmental Protection Agency. 1986. Ambient water quality criteria for bacteria. EPA440/5-84-002. http://www.epa.gov/waterscience/beaches/1986crit.pdf.

Environmental Protection Agency. 1989. Surface water treatment rule. http://www.epa.gov/OGWDW/therule.html#Surface

Environmental Protection Agency. 2001. Total coliform rule. www.epa.gov/safewater/therule.html#Total.

Environmental Protection Agency. 2004. Water Quality Standards for Coastal and Great Lakes Recreation Waters. www.epa.gov/fedrgstr/EPA-WATER/2004/November/Day-16/w25303.htm

Hartig, J. H., M. A. Zarull, T. M. Heidtke, and H. Shah. 1998. Implementing ecosystem-based management: lessons from the Great Lakes. Journal of Environmental Planning and Management 41:45-75.

Health and Welfare Canada. 1980. A study of disease incidence and recreational water quality in the Great Lakes. Phase 1. Publ. No. 81-EHD-67. Ottawa, ON: Environmental Health Directorate, Health Protection Branch.

Health Canada. 1995a. Notifiable disease summary. Canada Communicable Disease Report 21–18:166.

Health Canada. 1995b. Foodborne and waterborne disease in Canada. Annual Summaries. Ottawa, ON: Health Protection Branch.

Health Canada. 1995c. Great Lakes water and your health / Les eaux des Grands Lacs et votre sante. Ottawa, ON: Great Lakes Health Effects Program, Environmental Health Directorate, Health Protection Branch.

International Joint Commission. 1914. Progress report of the International Joint Commission on the reference by the United States and Canada in re the pollution of boundary waters whether or not such pollution extends across the boundary in contravention of the treaty of January 11, 1909 and if so, in what manner or by what means is it possible to prevent the same. www.ijc.org/php/publications/pdf/ID35.pdf

International Joint Commission. 1918. Final report of the International Joint Commission on the pollution of the boundary waters reference. www.ijc.org/php/publications/pdf/ID33.pdf

International Joint Commission. 2004. 12th biennial report on Great Lakes water quality. http://www.ijc.org/php/publications/html/12br/english/report/index.html

Myers, D. N., M. A. Thomas, J. W. Frey, S. J. Rheaume, and D. T. Button. 2000. Water Quality in the Lake Erie-Lake Saint Clair Drainages, Michigan, Ohio, Indiana, New York, and Pennsylvania, 1996–98. US Geological Survey Circular 1203.

Phelps, E. B. 1907. A method for calculating the number of *B.coli* from the results of dilution tests. Reports and Papers of the American Health Association 33: 9-13.

University of Wisconsin-Milwaukee. 2003. WATERbase Program. http://waterbase.glwi.uwm.edu/

4

Societal Changes and their Impact on Waterborne Infections in the United States and Canada: A Broad and Critical Perspective

S. A. Sattar and J. A. Tetro

© 2008 IWA Publishing. *Effective Cross-Border Monitoring Systems for Waterborne Microbial Pathogens*. Edited by J. B. Rose and E. A. Dreelin. ISBN: 9781843391692. Published by IWA Publishing, London, UK.

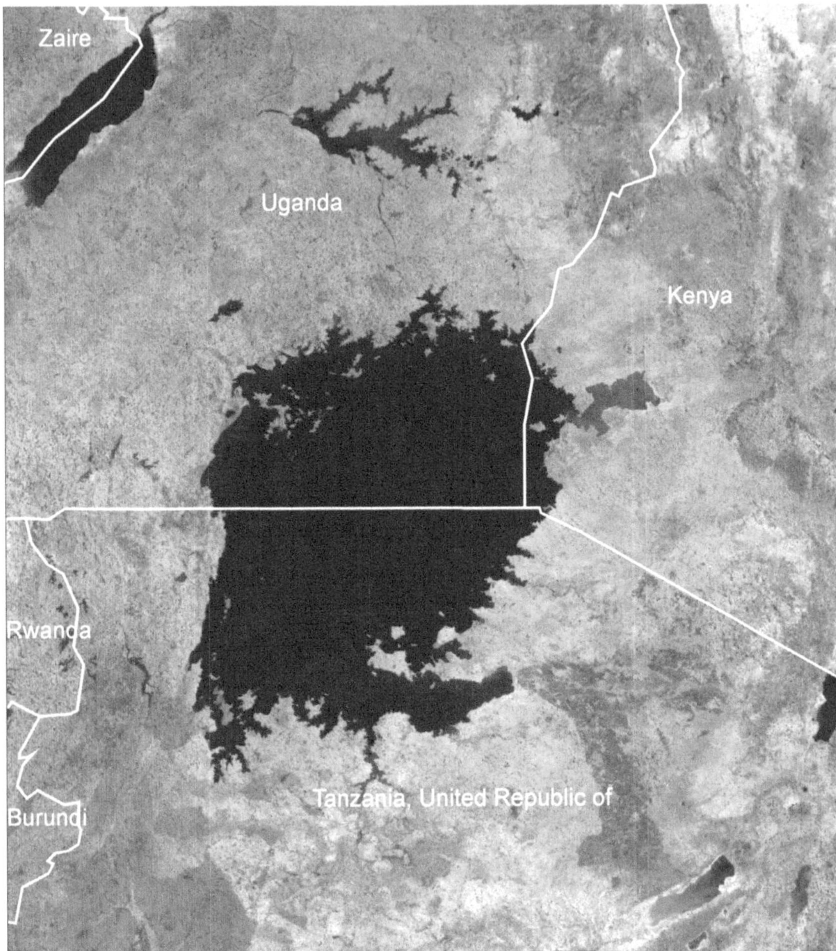

Lake Victoria. *Image: MSU Center for Global Change and Earth Observations*

"Human development and population growth exert many and diverse pressures on the quality and quantity of water resources and on access to them. Nowhere are the pressures felt so strongly as at the interface of water and human health."

- World Health Organization (2003)

4.1 INTRODUCTION

As is the nature of many human endeavors, 'progress' and 'development' spawn their own problems, some quite intractable (Macpherson 2005; Sattar *et al.* 1999; Smolinski *et al.* 2003). This chapter is a critical look at those societal factors that impact on the quality of fresh waters in this region in particular. The perspective is primarily that from human-health-related microbiology and is aimed at fostering a more reasoned understanding of the issues in the hope of promoting better public policies on management and use of our water resources.

Obviously, not all the societal factors given in Table 4.1 may have an equal bearing on the potential of water to spread infections. Nor should they be considered in isolation from issues such as climate change, which may or may not be anthropogenic. Of course, natural disasters, significant among them being, cyclones, earthquakes, hurricanes and tsunamis, are often difficult to predict, prevent and control. Nevertheless, mitigating such hazards has never been more urgent, especially in view of unabated human encroachment of vulnerable areas. This also blurs the lines between 'natural' and 'human-induced'.

We also consider it imprudent to exclude the rest of the world when discussing the health impact of water quality in the U.S. and Canada. The increasing closer links between various regions and peoples in today's world necessitate a less insular perspective, at least at the outset.

Additionally, a better appreciation of current water quality issues, and changes that may soon come, requires at least a cursory review of the recent past. Undeniably, our potable and recreational waters are a lot safer now than they were even a few decades ago. Nevertheless, there is still much room for improvement and also an urgent need to recognize and forestall problems that may arise from on-going societal changes.

Table 4.1. Categories of Anthropogenic and Natural Factors with Known or Potential Impacts on Waterborne Spread of Infectious Agents.

CATEGORY	FACTOR	TYPE OF IMPACT
DEMOGRAPHIC	Increasing life-expectancy	Aging process lowers general resistance to pathogens. Increasing numbers of seniors represent an expanding pool of susceptibles to opportunistic pathogens in water.
	Population movements	Increasing mixing of populations through travel and immigration. More frequent importations of enteric pathogens.
	Grouping of susceptibles	Increasing numbers of children and seniors in group settings. This facilitates spread of many enteric pathogens and enhances the burden on waste-handling facilities.
	Acquired immunosuppression	Diseases such as AIDS lower body's resistance to other pathogens and such individuals also excrete greater numbers of enteric and other pathogens into their surroundings.
URBANIZATION	Sewage effluents	Increasing volumes of liquid human wastes entering receiving waters.
	Changing drinking water treatment	Chlorine being replaced with other chemical and physical means to reduce formation of harmful disinfection by-products
	Biosolids	Increasing amounts of raw or partially-treated biosolids being applied on land with potential for ground and surface water contamination.
	Chemical pollutants	Exposure to waterborne chemicals that may compromise the immune system or disrupt hormone functions.
	Urban decay	Lack of renewal and expansion of facilities for water treatment as well as waste collection and disposal.

CATEGORY	FACTOR	TYPE OF IMPACT
	Pharmaceuticals	Long-term exposure to even low levels of pharmaceuticals may influence body's normal immune and hormonal functions.
	Microbicides	Certain types of common microbicides may persist in waters and may also add to the already serious problem of antibiotic resistance.
LIFESTYLE CHANGES	More frequent/faster travel	Facilitates the movement of human and animal enteric pathogens.
	Food importations	Greater varieties and quantities of enteric pathogens being introduced along with foods being imported from regions with higher levels of environmental pollution.
	Recreation	Increasing use of water for body-contact recreation. Biofilms in treated recreational waters pose health risks from opportunistic pathogens.
TECHNOLOGY	Induced immunosuppression	Organ transplantation and cancer therapy often entail suppression of the normal immune responses and thus increasing susceptibility of many types of pathogens.
	HVAC systems	Provide many opportunistic pathogens with suitable niches indoors thereby increasing human exposure.
	Hot water tanks	Opportunistic pathogens such as legionellae and environmental mycobacteria can readily grow in such settings and can be disseminated through aerosolization indoors.
AGRICULTURE AND ANIMAL HUSBANDRY	Agricultural wastes	Pathogens in manure spread on land may contaminate surface and ground waters. The risk of such contamination is higher in rural and peri-urban settings where no or minimal treatment may be given to potable waters.

CATEGORY	FACTOR	TYPE OF IMPACT
	Animal- and agri-wastes	The rising numbers of animal breeding/feeding operations adding to animal wastes entering the water environment. Zoonotic organisms such as *E. coli* O157:H7, *Cryptosporidium parvum* and hepatitis E virus increase the risk of human infections from water.
POWER GENERATION		Increasing quantities of water used for power generation may return otherwise unpolluted, but may increase the temperature of receiving waters and favor the growth of pathogens such as free-living protozoa.
	Dams	Construction of dams provides expanded habitats for secondary hosts for human pathogens and also increases areas where insect vectors such as mosquitoes can breed.
INDUSTRY	Mining	Large quantities of water are withdrawn for mining operation and wastes and tailings from such operations may impact more on the chemical quality of receiving waters.
	Manufacturing	Pulp and paper industry, for example, may discharge effluents with a high microbial content, including potential human pathogens.
CLIMATE CHANGE	Global warming	Climate change leading to higher levels of water pollution and also favoring breeding of insect vectors.
OTHER	Algal blooms	Algal growth in water may increase due to global warming and this could elevate the level of toxins; algae also harbor waterborne pathogens such as the one that causes cholera.

CATEGORY	FACTOR	TYPE OF IMPACT
	Natural disasters	Leading to increased levels of water pollution with simultaneous disruption of water treatment and supply systems.
	Political upheavals	Civil unrest or military action can disrupt the functioning of urban intra-structures such as water and waste treatment.
	Terrorism	Enhanced risk of accidental or deliberate release of infectious bioagents into water systems.

4.2 BACKGROUND

While water is generally recognized as essential for life, it is often treated as an inviolable and inexhaustible resource. As a result, water, in particular freshwater, has historically been at best mistreated and at worst egregiously abused. Modernization, as typified by industrialization, mechanized agriculture and urbanization, continues assaulting our vulnerable and fragile water resources with discharges of toxics and pathogens, thus threatening irreversible damage and the ensuing impact on human and ecosystem health. In other words, and quite paradoxically, many societal changes in the form of 'development' have direct and indirect impacts on the quality and safety of water.

The great plagues in the early days of urbanization forced the introduction of sewerage systems to simply collect and dispose of the community's liquid wastes into nearby receiving waters. The ensuing decline in the microbiological quality of water from these point sources precipitated fresh waves of enteric diseases such as cholera and typhoid.

In the industrial era that followed, increasing quantities of water were withdrawn by factories and mines for production and also to remove wastes, which were often quite toxic, for disposal in the nearest watershed. The prevailing philosophy that 'dilution is the best solution to pollution' was soon squarely debunked, but without convincing the society at large that protecting and conserving water resources should be a first priority in societal advancement. Unfortunately, the residual effects of the times past continue to be felt even to this day. Mercury pollution of water is a good example here (UNEP 2003).

The need for safe water to maintain health and hygiene should be an overriding factor in the development of environmental policy, but competing pressures from numerous other quality-of-life issues often usurp this consideration, thus increasing the risk of spread of infections from water used for drinking and body-contact recreation. Admittedly, large outbreaks of waterborne infection in Canada and the U.S. occur comparatively infrequently. But, the ones that have been recorded in about the past ten years (MacKenzie *et al.* 1994, Hrudey *et al.* 2003, Bowie *et al.* 2003) are a harbinger for what may come.

4.3 INFECTIOUS DISEASE SPREAD BY WATER

Globally, infectious diseases cause >32% (18 million) of the 56 million deaths/year (WHO 2002). This is obviously an underestimation because, unlike traffic accidents, murders or drownings, for instance, many natural causes of death are not, or cannot be, investigated sufficiently for a possible underlying infectious basis. Moreover, and perhaps more importantly, infectious agents are increasingly being linked to disease conditions and disorders where their role was never suspected; hardening of arteries leading to fatal heart attacks (Ismail *et al.* 1999) and terminal cancers are just two examples here (Goedert 2005). Therefore, the true human health impact of infections is most likely far greater than presently realized (Ewald 2004). With regards to waterborne infections in today's world, precise figures remain unavailable, but the data from the WHO (2002) show that globally, unsafe water causes >3.3 billion cases of illness and 5-10 million deaths/year.

4.4 WATERBORNE INFECTIONS IN U.S. AND CANADA

In this region, drinking water-associated outbreaks of once common diseases such as typhoid and cholera have been replaced, at least in part, by those of pathogens which were not known to spread through potable water. The recent and fairly large outbreaks due to *Toxoplasma gondii* (Dubey, 2004; Slifko *et al.* 2000), *Cryptosporidium parvum* (Smith and Rose 1990) and *Escherichia coli* O157:H7 (Hunter 2003) are just three examples here. Each such outbreak also raises more questions about the ability of our current water treatment methods to deal with known and emerging pathogens (Leclerc *et al.* 2002; Nwachcuku and Gerba 2004). Further, the human health impact of biofilm-based organisms in American and Canadian drinking water supplies is just beginning to be explored (Falkinham 2001, Anderson 2002).

In Canada, surveillance for waterborne outbreaks of infectious disease was initiated only in 1974. Between that date and 1996, the latest year with available data, over 200 drinking water-related outbreaks were recorded, with at least 8,000 confirmed cases (Environment Canada 2003). Studies by Payment *et al.* (1991) suggest that, at least in certain parts of Canada, municipally-supplied tap water may be responsible for nearly 35% of all waterborne gastrointestinal illnesses. Estimates by Environment Canada (2003) indicate that there are roughly 90,000 cases of waterborne illness and about 90 deaths in Canada annually.

In the U.S., between 1971 and 2000, there were 730 outbreaks of infection from contaminated drinking water (Lee *et al.* 2002). Figure 4.1 summarizes the data for such outbreaks for 1991-2000; in the last two years of that period there appeared to be a reversal of the downward trend recorded in previous years.

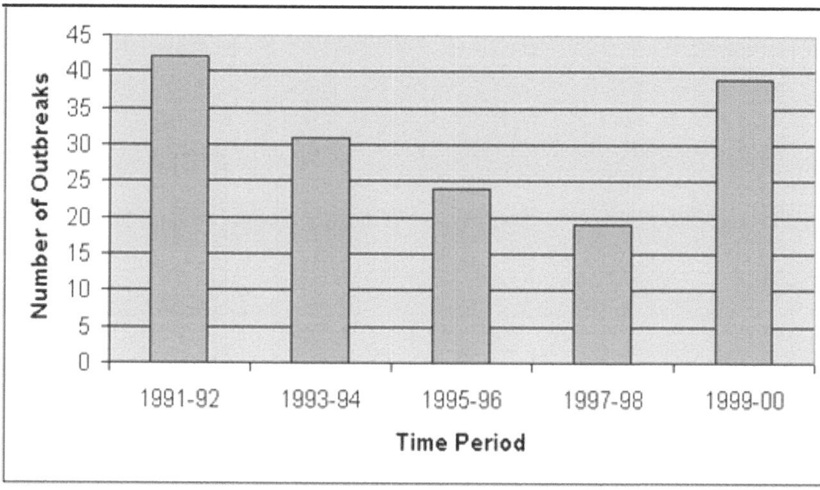

Figure 4.1: Number of Drinking Waterborne Outbreaks in the United States, 1991-2000. (Based on data from Lee *et al.* 2002)

The role of recreational waters in the spread of infections is even less well-defined (Castor and Beach 2004) and Canadian data is particularly scarce on this topic. But, if the frequency of beach closures is an indicator of the extent of fecal pollution of coastal waters, then, in Canada, there were over 2,800 such closures during 1992 and 1995 (Environment Canada 2003). As can be seen from Figure 4.2, there is quite a dramatic upward trend in the number outbreaks related to untreated recreational waters in the U.S. (Craun *et al.* 2005). The wide variety of pathogens incriminated in outbreaks from untreated recreational waters is illustrated in Figure 4.3. The perennial question of any increases recorded are due to better surveillance or an elevation in the actual number of cases is just as difficult to answer here as it is with regards to other types of vehicles.

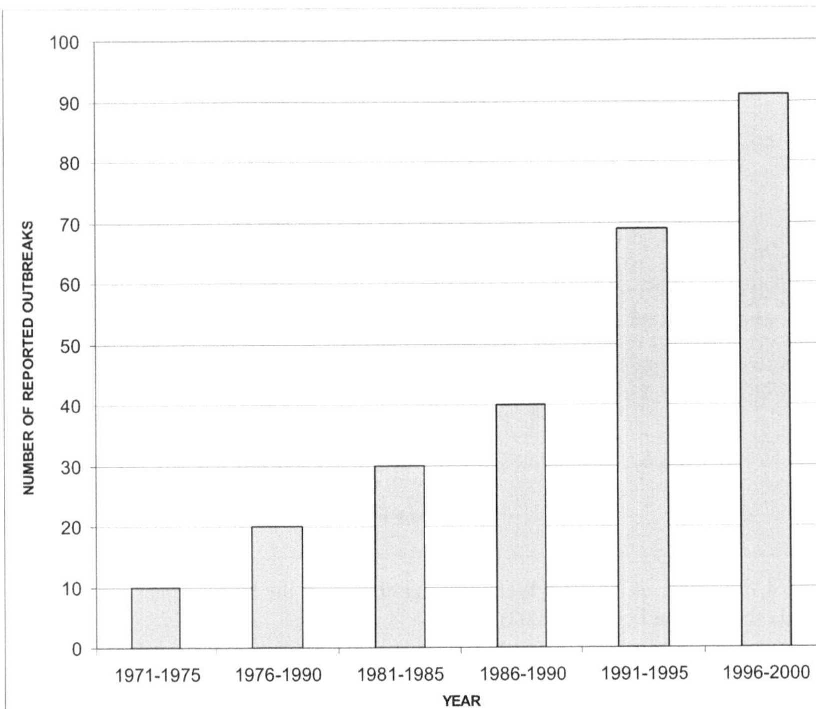

Figure 4.2. Outbreaks associated with recreational waters in the United States, 1971-2000. (Modified from Craun *et al.* 2005)

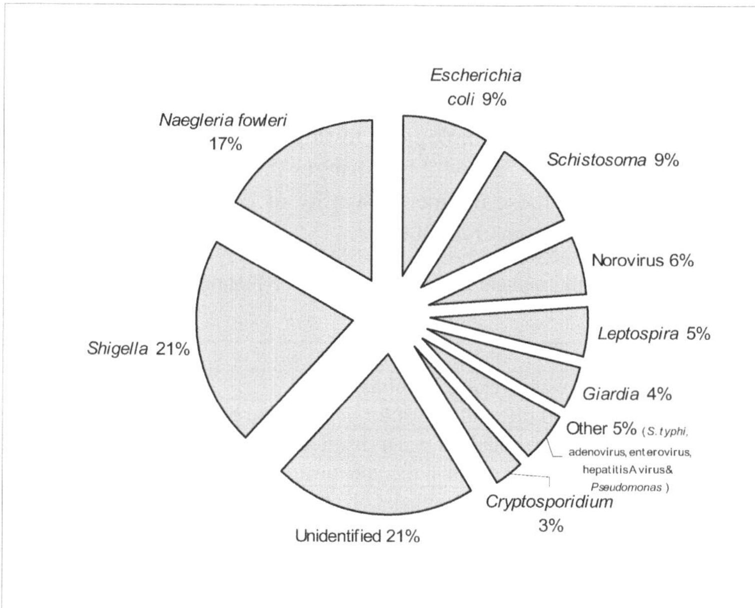

Figure 4.3. Etiology of Outbreaks Associated with Untreated Recreational Water in the U.S., 1971-2000 (Modified from Craun *et al* 2005).

Another, and perhaps the least well-appreciated, mode of spread of infections in the U.S. and Canada is through the inhalation of aerosolized water (Falkinham 2003). This is a modern-day phenomenon to a very large extent because of the wide-spread use of engineered environments such as heating, ventilation and air-conditioning (HVAC) systems, as well as hot water supplies that provide highly suitable indoor niches to many opportunistic pathogens. *Legionella pneumophila*, the main cause of community-acquired pneumonia, and environmental mycobacteria, being incriminated in a variety of respiratory, enteric and skin diseases, are good examples of such opportunistic pathogens (Falkinham 2002).

Environmental mycobacteria are worthy of additional comments here because of their significance as emerging waterborne pathogens in this region (Falkinham 2002; Wagner 2004; Wallace *et al.* 1998). Many such species are well-recognized components of water-related biofilms; they can be isolated from even small volumes of recreational waters and are also very common in municipal water distribution systems (Hunter *et al.* 2001; September *et al.* 2004). 'Lifeguard lung' (Rose *et al.* 1998, Angenent *et al.* 2005) and 'hot-tubber lung' (Marras 2005) are among the recently described clinical conditions in otherwise healthy persons exposed to aerosols from *treated* recreational waters and those used for showering (Falkinham 2003; Marras and Daley 2002). Further, recent evidence shows that changes in water disinfection regimes to reduce the formation of trihalomethanes may actually increase the proliferation of mycobacteria (Pryor *et al.* 2004). The true extent of this remains unknown simply because routine microbiological testing of water does not include detection of mycobacteria. Table 4.2 summarizes the features of environmental mycobacteria that make these organisms worthy of consideration as emerging waterborne pathogens (Primm *et al.* 2004).

Table 4.2. Human Health Impacts of Environmental or Nontuberculous Mycobacteria (Primm *et al.* 2004)

Trait
Common components of water-related biofilms
Many are slow-growing and difficult to test for on a routine basis
Generally more resistant to residuals of water disinfectants in municipal systems
Often difficult to treat with conventional therapeutic agents
Can be acquired through ingestion, inhalation, injection and also by contact with broken skin
May take several months or longer for clinical manifestation of infection
Cause diseases not only in the immunosuppressed but also in apparently healthy individuals
Often associated with free-living protozoa which afford them further protection

There is also increasing concern on the health impact of simultaneous or sequential exposure to chemicals and pathogens (Lewis *et al.* 2002, Germolec 2004). Table 4.3 gives examples of those environmental chemicals that may enhance human susceptibility to infections.

As summarized in this section, waterborne spread of infections remains a concern for human health in this region. Even though many once common enteric pathogens are no longer a threat, others are being recognized as filling the void. This, together with a variety of on-going changes, as described below, is raising the specter of newer and potentially harder-to-deal-with issues.

Table 4.3. Environmental chemicals with potential for altering resistance of humans to infectious agents.

Chemical	**Comments**
POLYCHLORINATED BIPHENYLS (PCBs)	Till 1979, PCBs were used for many purposes in the U.S. and Canada (Environment Canada, 1996). The use of PCB's was phased out due to their ability to bioaccumulate and cause immune dysfunction and mutations in experimental animals. While such bioaccumulation has also been documented in humans (Juan *et al.* 2002; Bloom *et al.* 2005), we are unaware of and clear links between it and lowered resistance to infectious agents.
DIOXINS	While they are considered among the most potent anthropogenic carcinogens (Steenland *et al.* 2004), there is no established link between them and enhanced susceptibility to infections.
MICROBIAL MUTAGENS	Such chemicals can impact on human health by causing enhanced mutations in microorganisms. Textile dyes are an example here (Schneider *et al.* 2004).
ENDOCRINE DISRUPTOR CHEMICALS (EDCs)	More recently, certain types of chemicals have been recognized for their ability to disrupt endocrine function in aquatic animals and wildlife (Lathers, 2002), resulting in abnormal development. EDCs are now under intense investigation to determine their impact on human health (U.S. EPA, 1998). .
PESTICIDE RUNOFF	Pesticides are almost a requirement of agriculture in the U.S. and Canada. Unlike antimicrobials, pesticides are less likely to cause resistance in bacteria, but, much like environmental toxins from industrial sources, pesticides may cause immunosuppression (Vial *et al.* 1996; Choi *et al.* 2004; Wilson *et al.* 2004).

4.5 SOCIETAL CHANGES

Of the factors given in Table 4.1, immunosuppression and drug resistance were virtually unknown some 30 years ago. Today, they are among the major issues impacting our health in general and, more specifically, the role of water in spreading infections (Falkinham 2003; Rusin *et al.* 1997). Other factors, e.g., natural disasters (Schneider *et al.* 1997), climate change (Charron *et al.* 2004; Rose *et al.* 2001; Weiss and McMichael 2004) and urban decay (Wallace and Wallace 1996) can also enhance, directly or indirectly, the potential of water to spread infections.

The combined effect of a variety of societal changes has been responsible for a significant rise in water usage (Figure 4.4). The very simple reality here is that the larger the volumes withdrawn the higher are the quantities of spent waters with toxics and microbial pathogens that enter our aquatic systems. In Canada, the magnitude of increase in water withdrawals has yet to surpass the amount of renewable water resources, but certain parts of the U.S. have already been experiencing serious water shortages (Beuhler 2003). The situation is much worse in other parts of the world. According to the United Nations, the number of countries experiencing serious water shortages continues to rise; currently nearly 20% of the world population in 31 countries suffers from water shortages and that number is expected to nearly double in the next 50 years (UN 2003).

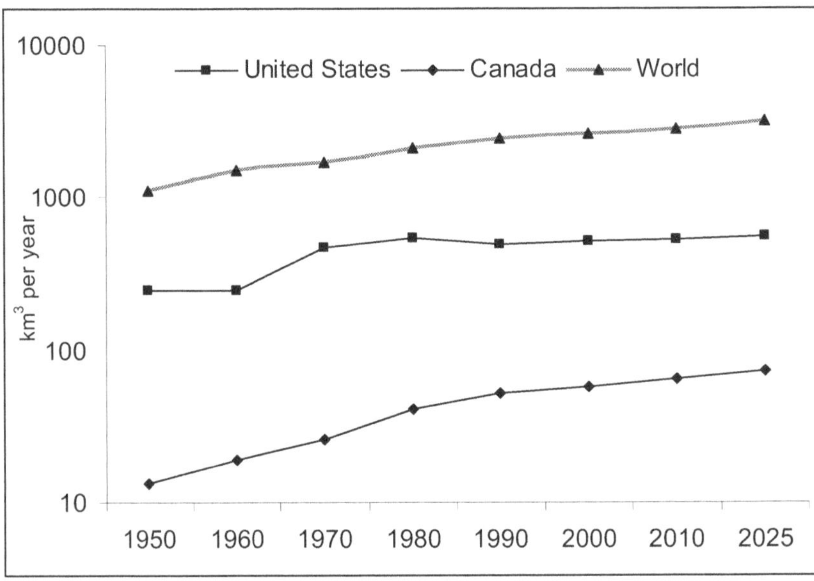

Figure 4.4. Actual and estimated freshwater withdrawals, 1950-2025 (Shiklomanov 1999)

To understand the contribution of societal changes to waterborne spread of infections one must also look closely at how water is used today. In the U.S., water withdrawals for various needs increased in the past 50 years from 175 billion gallons (665 billion L)/day to 323 billion gallons (1,227 billion L)/day. Most of this is freshwater (262 billion gallons/995.6 billion L), of which over half (135 billion gallon or 513 billion L) is devoted strictly to the generation of thermoelectric power. In contrast, only 27.3 billion gallons (104 billion L) are for public consumption. Agriculture and aquaculture use 95 billion gallons (360 billion L)/day while the rest is used for other industrial activities, including mining.

In Canada, the picture is quite different. Of the 10.7 billion gallons (40.6 billion L) of freshwater used/day in 1996, almost 70% was for thermoelectric power. Other manufacturing processes used 13.5% while mining used only

1.7%. Municipal and agricultural use only accounted for 12.8% and 2.6%, respectively. Rural use accounted for approximately 2%.

When one looks at population growth, a correlation with water consumption was present until the 1980s, when a better awareness for water conservation started to set in (Environment Canada 1996). From that point on, the correlation ended and by the year 2000, water consumption rates had all but leveled off. Doubtless, this is in large part due to successful public education programs by governmental and non-governmental organizations alike.

The increasing withdrawals of freshwater mentioned above entails that ever-larger volumes of polluted waters are being 'returned' to the environment. Concurrently, there have been several technological developments in an effort to treat water and wastewater. The first of these was drinking water treatment/disinfection; its rapidly widening application saw a dramatic drop in the numbers of cases of many common enteric infections (Sargeant 2005). Next was the development of wastewater treatment, which removes a large majority of enteric and other pathogens either through physical and/or chemical means (Godfree and Farrell 2005). Yet, and in spite of a reduced pathogen load, the effluents returned to the environment can be a cause for concern. For instance, recent studies have identified them as regular and increasing significant sources of antibiotics and other pharmaceuticals into the water environment; although the levels of such pollutants in receiving waters are believed to be too low to be of human health concern in the U.S. and Canada (Miao *et al.* 2004; Schwab *et al.* 2005), we believe that chronic exposure to such anthropogenic chemicals, even in low doses, should be a cause for concern, particularly in view of their potential to generate drug resistance and interfere with body's defense mechanisms.

While there have been major improvements in municipal water supply and distribution systems, the same cannot be said about rural water supplies. And it is this area that is likely to exert a greater impact on human health as the population in peri-urban areas increases. The U.S. and Canada are no exceptions in this context. Corkal *et al.* (2004) have recently enunciated the impact of water quality on the lives of nearly 4 million Canadians living in rural settings.

Because of a much larger base here, the commercial/industrial sector is often blamed for a negative impact on water quality in this region and beyond. While this may be true for chemical pollutants, especially when major spills occur, this sector has remained much less significant as a contributor to microbial pollution of our water resources. However, there are emerging concerns as summarized below.

4.5.1 Population-Related Factors

The continuing change in the demographics in the U. S. and Canada is leading to an increased potential for the spread of infectious disease through water. Many of these factors may have only an indirect relationship with the quality of water but as these trends continue to grow, the potential for problems will as well.

4.5.1.1 Demographics

To understand the potential for watershed contamination by municipal activities, one must know the current spectrum of the population within these communities and how they are changing. The urban community is a different place than it was 50 years ago and although we have seen much advancement in reduction of industrial pollutants, municipal-based pollution seems to have taken over as the primary environmental concern. Worse, there may be no easy and widely-applicable solution to this problem. Nearly all urban settings continue to be substantial sources of microbial and chemical pollution from storm waters they discharge.

Average life expectancy in the U.S. and Canada is higher than ever and is still increasing, making seniors the fastest growing portion of the population (U.S. Census Bureau 2004). By 2041 close to 25% of Canadians will be aged 65 or over (Health Canada 2002); the corresponding figure in the U.S. will be about 20% (U.S. Census Bureau 2004). This is in marked contrast to 1951 where seniors accounted for <10% of the total population in both these countries.

The general senescence of the body's defense mechanisms in the elderly increases their susceptibility to many infectious agents, making them increasingly important sources of pathogens for others in their surroundings (Stevenson and Loeb 2004). Further, more frequent and possibly inappropriate use of antibiotics in homes for the aged increases the risk of antibiotic resistance and enhanced seeding of the water environment with such organisms.

4.5.1.2 Urbanization

While increasing urbanization is not peculiar to this part of the world, it is nonetheless a significant factor in enhancing the levels of microbial and chemical pollution of our surface as well as ground waters and its ensuing negative impacts on human health. Amongst the largest cities in the U.S., only New York has seen a population decrease in the last 20 years (Table 4.4). The largest increase has occurred in Atlanta, GA, where the population has gone up by some 85% (Kolankiewicz and Beck 2000). In Canada, the urban population tends to grow some 1.5% every year, a rate that cannot be explained simply by the natural rate of population growth (Simmons and Bourne 2003). Thus, there is a movement towards urban centers along with the impact of new arrivals from outside the country.

Table 4.4. Top ten urbanized cities in the U.S. and the change in their urban sprawl and population (Kolankiewicz and Beck, 2000)

	Urbanized Area	**% growth in per capita land consumption**	**% growth in population**
1.	Atlanta, GA	42%	84%
2.	Houston, TX	26%	73%
3.	New York City, NY-NJ	24%	-1%
4.	Washington, DC-MD-VA	41%	36%
5.	Philadelphia, PA-NJ	48%	5%
6.	Los Angeles, CA	-8%	37%
7.	Dallas-Fort Worth, TX	-15%	59%
8.	Tampa-St. Petersburg, Clearwater, FL	13%	98%
9.	Phoenix, AZ	-18	132%
10.	Minneapolis-St. Paul, MN	21%	22%

4.5.1.3 Grouping of Susceptibles

Susceptible individuals can be defined as having either a naïve immune system or one that is debilitated due to age, drug/radiation therapy or infections such as AIDS. The major concern with the rising number of susceptibles is the potential for the colonization and evolution of microbes within the host.

As the number of the elderly increases in the community, so does the population of nursing homes and long-term care facilities (LTCF). In 1974, less than a million Americans $\geq$65 were in a nursing home. That has increased to nearly 1.5 million in 1999. In Canada, as of 1996, >500 thousand elderly were in various LTCF (Health Canada 1999). With such a large number of people living in a grouped setting, one can well imagine the burden on the water system in the event of an outbreak.

Apart from the elderly, children are perhaps the largest group of susceptibles. And, as a result of daycare and other child-centered activities, children are continually grouped together in ever-larger numbers. The number of American children under 5 years of age cared for in daycare centers rose from 5 million to 10 million from 1985 to 1999 (U.S. Census Bureau 2003). That number has most likely increased further over the last 6 years. When an infectious disease strikes even one individual in an environment such as a children's daycare centre, much like in an LTCF, the potential for an outbreak is high (Sattar *et al.* 1999). Outbreaks in this group are almost endemic and include many gastrointestinal infections (Ford-Jones *et al.* 2000).

Another relatively new but important group of susceptibles are those suffering from AIDS. Since the early 1980's the disease has struck over 900,000 Americans (Fleming *et al.* 2002) and over 56,000 Canadians (Health Canada 2003). AIDS almost completely abrogates the immune system, thus enhancing vulnerability to even those infectious agents which may be harmless to otherwise healthy individuals. Significant among these is waterborne *Mycobacterium avium* (Falkinham *et al.* 2001). In the case of *Cryptosporidium parvum*, another important waterborne pathogen (Crawford and Vermund 1988), there is speculation that it evolved from a benign to a more pathogenic form due to its persistence in AIDS cases (Pieniazek *et al.* 1999).

Cancer patients and organ recipients also fall into the category of susceptibles. Chemotherapy and radiation therapy are both known to reduce immune function and render a patient weaker in the fight against infections (Gea-Banacloche *et al.* 2004). The impact of emerging waterborne pathogens on the health of those in healthcare settings has been identified (Emmerson 2001).

4.5.1.4 Foodborne Diseases

Although not directly related to water, the mounting influx of imported foods from areas with lower standards of hygiene has the potential to affect the microbiological quality of water in the U.S. and Canada. Several emerging pathogens have entered these two countries through foods contaminated at source. For example, in 1996, *Cyclospora cayetanensis*, a relatively unknown parasite in North America, but endemic in Central America, caused an outbreak involving raspberries that affected 978 Americans over 20 States. Since then, there have been ten more outbreaks leading to over 3000 cases of illness (Mansfield and Gajadhar 2004). The shedding of the parasite in the feces of patients can contaminate waters and could make the parasite endemic in this region.

Hepatitis A virus (HAV) has also caused several foodborne outbreaks in the U.S. (Rosenblum *et al.* 1990; Niu *et al.* 1992; Hutin *et al.* 1999). As with *C. cayetanensis*, HAV is able to survive the varying environmental conditions of food transportation and also remain viable in liquid and solid municipal wastes with the potential for contamination of receiving waters. Thus, an increase in the levels of such pathogens in the water environment enhances the risk of infections from them.

4.5.2 Municipal Factors

Although municipal uses account for less then 15% of all freshwater withdrawals in the U.S. and Canada, there is no doubt that the impact of water use in this sector is by far the most important to human health.

4.5.2.1 Municipal Waterworks

With an increase in population comes an increased burden on the municipal water systems. Most large cities and several smaller towns now have water treatment facilities employing a multiple-barrier approach to ensure drinking water safety. For most systems, there is at least one disinfection step. Chlorine continues to be the predominant chemical for disinfection regardless of worries of byproducts that may be potential carcinogens. There is a great debate over trihalomethanes but most believe that the benefits of chlorination far outweigh the possibilities of not using chlorine should no other disinfection option be possible. Regardless, evidence that pathogens such as *Cryptosporidium* may still evade the treatment process (Aragon *et al.* 2003) has been forcing a review of basic water treatment processes.

As water finds its way from the treatment plant to the tap, it is not entirely immune to certain microbial assaults. While encroachment of contamination through leaks in the system may occur from time to time, it is primarily from biofilm build-up on the inside surfaces of pipes in the network. Such biofilms can cause corrosion, taste and odor problems, deplete disinfectant residuals and also continually release a variety of opportunistic pathogens into the otherwise treated water. Biofilm-based pathogens are currently a large area of study (Ghannoum and O'Toole 2004). Of note is *Mycobacterium avium*, which represents a major threat to immunocompromised individuals (Aronson *et al.* 1999). Other potential biofilm-based bacteria include *Pseudomonas* (hospital acquired infections), *Moraxella* (ocular and respiratory infections), *Aeromonas* (urinary tract disorders) and *Legionella* (Rusin *et al.* 1997; Pryor *et al.* 2004. Norton and LeChevallier (2000) have determined that the type of pipe used (i.e. iron vs. polyvinyl chloride) was important in determining the microbial mix in water distribution system biofilms. While they did not test for mycobacteria, another U.S.-based study has found mycobacteria in water distribution systems with levels ranging between 10 and 700,000 colony forming units (CFU)/L (Falkinham *et al.* 2001). As mentioned above and also summarized in Table 4.2, water biofilm-related environmental mycobacteria must be considered important among emerging waterborne pathogens because many on-going societal changes increase our exposure and vulnerability to them.

The mounting concerns on the quality of potable waters have triggered a debate on the difference between 'tap water' and 'drinking water' (Beuhler 2002); considered synonymous till recently. The perceived or real fears that municipally-treated drinking waters may be harmful to human health, is behind the increasingly common use of pre-packaged, bottled or spring water as well as point-of-use (POU) water treatment devices in the U.S. and Canada (Ahmedna *et al.* 2004). At present, the pre-packaged-water industry is essentially self-regulated in Canada (Canadian Food Inspection System, 2003) and that in the U.S. is subject to FDA regulations (U.S. FDA 2003), but that the microbiological quality of such products may not be as good as generally perceived, and thus bottled water safety is a matter of some concern (Ahmedna *et al.* 2004; Chaidez and Gerba 2004). POU treatments also vary considerably in their basic design and the label claims they make (Canadian Food Inspection System 2003); unless they are maintained and used properly, they may generate a false sense of security with regards to the microbiological quality of water (Chaidez and Gerba 2004).

4.5.2.2 Wastewaters

On the other side of the water use spectrum is the generation of wastewater. Many large cities now have some form of wastewater treatment to reduce the discharge of potential microbial and chemical contaminants into the watershed. These large treatment plants are for the most part successful as they render sewage effluents relatively clean through a multi-stage process. For those that do not use a tertiary process, however, chemical pollutants that are not retained within the solids are released back into the environment, albeit at lower concentrations. Among the chemicals that may escape the treatment systems are common types of pharmaceuticals and microbicides and some of them are the target of current studies in the Great Lakes region (Jasim 2004).

4.5.2.3 Pharmaceuticals

Until recently, nearly 1 in 3 Americans was prescribed an antibiotic without a valid diagnosis (Nyquist *et al.* 1998), but extensive educational campaigns aimed at the general public and professionals alike may be improving the situation as a recent survey found no antibiotics among the top 10 prescriptions purchased in the U.S. (Stagnitti 2003).

Of those who require prescriptions, the elderly represent the highest number. According to Stagnitti (2003), from 1997 to 2000 the purchase of outpatient prescription drugs increased from U.S. \$72.3 billion to \$103 billion. The average expenditure for a person over 65 was \$1,102.00 in 2000, over double that of persons under 65. Not surprisingly, there were also twice as many prescriptions for those 65 and older compared to those under 65. Even so, the number of people taking prescription medication continues to be on the rise.

Antibiotics and other anti-infectives in receiving waters may confer higher levels of resistance or cross-resistance to antimicrobials and further contribute to the already serious problem of antibiotic resistance. As shown in Table 4.5, bacteria with such resistance have been found in waters receiving feces from non-point sources (Hagedorn *et al.* 1999). Antibiotic resistance profiles in such bacteria correlate well with those isolated from humans and animals in the surroundings; such pattern of antibiotic resistance may help in identifying the generally difficult-to-locate non-point sources of fecal contamination and also to assess the outcomes of clean-up measures instituted. Further, Yang and Carlson (2003) have shown that antimicrobial contamination of the Poudre River in Colorado is dependent on either agricultural runoff or wastewater influx (Figure 4.5).

Table 4.5. Prevalence of antibiotic resistance in fecal streptococci in Page Brook in Clarke County, VA (Hagedorn *et al.* 1999)

Drug and conc. (μg/mL)	% of resistant isolates from each source[a]					
	Beef cows (*n* = 1,398)	Chickens (*n* = 824)	Dairy cows (*n* = 728)	Deer (*n* = 1,245)	Humans (*n* = 1,579)	Waterfowl (*n* = 1,284)
Chlortetracycline						
40	1	90	17	0	78	1
60	0	68	8	0	39	0
Erythromycin						
7	1	89	4	0	90	1
15	0	77	0	0	76	0
Neomycin						
10	2	100	100	0	54	0
40	0	98	88	0	4	0
Oxytetracycline						
10	90	100	100	38	98	97
30	80	100	88	0	84	45
60	12	100	81	0	56	5
100	0	95	76	0	14	0
Streptomycin						
10	96	100	100	59	100	97
30	73	96	96	2	96	79
45	25	89	50	2	91	31
60	3	73	20	0	82	12
100	1	65	1	0	25	0
Tetracycline						
15	81	100	80	0	87	2
30	13	100	79	0	81	1
100	0	98	67	0	12	0

[a] n, total number of isolates from each source.

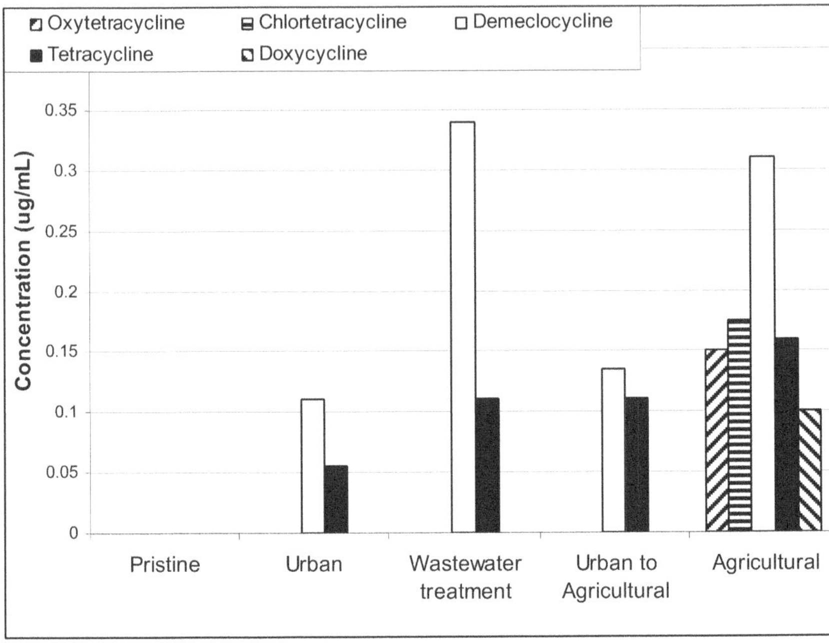

Figure 4.5. Occurrence of Tetracycline Residues in the Poudre River, Colorado (Yang and Carlson, 2003)

4.5.2.4 Microbicide Use

Apart from antibiotics, disinfectants, detergents and antiseptics and other personal care products are widely used. In 2002, Sanitary Supply Distributor Sales for these products totaled over 2 billion dollars in the U. S. alone (International Sanitary Supply Assoc. 2004). Many of these uses are for environmental cleaning ard thus discharged directly into sewers. As Patterson (2001) points out, while there may be regulations on the quality of wastewater being released into the watershed, there is no control over the amount of disinfectants, detergents and antiseptics used in the domestic setting. For example, the use of laundry detergent traditionally increased the phosphorus load affecting the quality of the final effluent product. With recent advances in laundry detergent technology, phosphates have been reduced or eliminated from these products. However, other chemicals, such as salts, which, like phosphorus, cannot be taken out of wastewater due to their solubility, are relatively high in the final effluent (Patterson 2001). Kummerer (2001) has shown that quaternary ammonium compounds (QAC) have relatively poor sorption rates promoting their discharge into the watershed.

As many active ingredients in disinfectants and antiseptics have unknown mechanisms of action (Russell 2002), the potential of low levels of microbicides to cause antimicrobial resistance is suspected but its impact remains to be assessed. However, lab-based studies have shown that bacteria with resistance to triclosan, a common microbicide in many personal care products, can also become cross-resistant to antibiotics with the same site of action (Levy *et al.* 1999). Traces of triclosan have already been detected in groundwaters in North America (Reiss *et al.* 2002).

4.5.2.5 Environmental Toxins

Many such anthropogenic chemicals entering receiving waters have the potential to alter immune function or interfere with hormonal activity, thus predisposing to frank and opportunistic pathogens. As mentioned above, there is also mounting concern on the combined deleterious effects of environmental chemicals and microorganisms on the health of humans (Lewis 2002; see Table 4.3)

4.5.2.6 Land Applications of Municipal Solid Wastes

The threat to water through landfills is perhaps the most frustrating problem that has emerged in the past few decades. Initially determined to be a safe environment in which to place unwanted solid wastes, the appearance of chemical contamination through leachates and plumes has all but destroyed some communities. Yildiz *et al.* (2004) looked at the now closed Keele Valley Landfill in Ontario and determined the quality and quantity of chemical leachates that can pose health hazards to groundwater sourced communities.

In terms of microbial hazards from landfills, Sobsey (1978) showed that enteric viruses were barely able to withstand the environmental pressures of landfills and that contamination through leachates was, as in the case of biosolids, quite remote.

The advancements in municipal wastewater treatment are leading to ever-increasing amounts of highly nutrient-rich biosolids, a processed form of sewage sludge. This form of solid municipal waste can be used to augment nutrients in farmed soil, provide landfill cover, produce topsoil or help in the reclamation of mining territories. Indeed, the land application of biosolids has become a large, if controversial, business (Leblanc *et al.* 2004) and is likely to increase because of the cost savings that can be realized by municipalities. Unless properly treated, handled, stored and applied, the very nature of biosolids can make them sources of chemical pollutants and microbial pathogens.

Gerba and Smith (2005) have shown that some 150 enteric pathogens may exist and thrive in raw sludge. Additionally, there is the presence of various chemical pollutants, such as antimicrobials, microbicides and heavy metals which may also pose an environmental risk. Depending on the process by which sludge is converted into applicable biosolids, microbial and organic chemical loads may decline. However, there is always a risk for heavy metal and residual contamination through either leaching or runoff. Furthermore, this risk may be heightened due to potential interactions of pathogens and irritant chemicals (Lewis *et al.* 2002). Many communities now have programs to prevent potentially harmful chemicals from entering the municipal waste stream, thus reducing contaminant loadings of soils and receiving waters.

Adequate treatment of municipal sludge, its safe storage, proper selection of the land disposal site as well as the right application ratios and disposal techniques are all essential to optimize the recycling of biosolids while virtually eliminating any human health risks from pathogens in them (Gale 2005). While regulations to ensure their quality and application practices are evolving, biosolids continue to have a poor public image.

4.5.3 Agricultural and Animal Husbandry

In the U.S., 52.6 million acres (21.3 million hectares) of land required irrigation in 2003 (National Agricultural Statistics Service 2004); according to (Coote and Gergorich 2000), the number in Canada was nearly 1 million acres in 1996 (0.4 million hectares). On average, the annual amount of water used/acre (0.4 hectare) was about 2 acre feet (2,466 m^3) or 650 thousand gallons (2.4 million litres) for both countries (National Agricultural Statistics Service 2004). The consumptive use of water, unlike both manufacturing and thermoelectric uses, is rather high at around 60% (Hutson *et al.* 2004). Thus, only about 30% of the water is actually returned to the watershed. That amount, however, does not come back unpolluted.

Water given to livestock is eventually returned to the environment as body fluids. Manure can be spread on crop fields as fertilizer for future planting seasons. The water from manure then has the ability to leach through soil or runoff through climatic erosion such as rain and snow melt. In irrigation, water will initially absorb into the soil and be consumed by the crops. The remainder will follow a similar route of leaching or runoff. In either case, the water is anything but clean and safe.

Manure by its nature is high in microbial as well as nutrient load and may contain pathogens such as the now infamous *Escherichia coli* O157:H7 and/or *Cryptosporidium parvum*. When runoff occurs either through overloading of manure or climatic events, pathogens can flow into surface waters or be preferentially transported through the overburden to groundwaters, either of which may be sources of drinking water. This chain of events was responsible for a now well-known outbreak of drinking water-related illness in Canada (Health Canada 2000).

The Walkerton incident (Hrudy *et al.* 2003) may have been the most publicized outbreak due to runoff, but the incriminated types of pathogens are most likely not uncommon in other watersheds receiving agricultural wastes. Stanley and Jones (2003) have shown that sheep and cattle harbor *Campylobacter* and the application of manure is a source of environmental contamination. Johnson *et al.* (2003) found both *E. coli* O157:H7 as well as *Salmonella* spp. in an Alberta watershed proximal to an agricultural site. According to Sischo *et al.* (2000), dairy farms had over 90% prevalence of *Cryptosporidium*, a parasite known for having caused several large waterborne outbreaks in Milwaukee, Wisconsin (1993) and North Battleford, Saskatchewan (2001). Based on probability studies, Sischo *et al.* (2000) demonstrated that the likelihood of water contamination with *Cryptosporidium* was positively correlated with manure spreading. As if to corroborate, a public inquiry suggested that runoff due to spreading may have been the cause of the North Battleford outbreak, which infected up to 7,000 people (Saskatchewan Justice 2002) and there is wide speculation that runoff was also the cause of the Milwaukee incident, which resulted in over 400,000 cases of the infection (Mackenzie *et al.* 1994).

The more recent trend in livestock breeding/feeding by concentrating large numbers of animals into a small farming area may also impact on water quality (Kirkorn 2002). A Concentrated Animal Feeding Operation (CAFO) is described as containing more than 1,000 animal units; or in terms of swine, 2,500 head. There were over 238,000 such CAFO farms in the U. S. in 2002 totaling annually 500 million tons (227 million kg) of manure. This compares to human waste estimate of about 150 million tons (68.1 million kg) for the entire American population (U.S. EPA 2003). Unlike effluents from municipal sewage treatment facilities, animal wastes often enter receiving waters as non-point sources, thus creating a substantial challenge for detection and control.

Several pathogens have been identified from the manure of CAFO animals (Guan and Holley 2003). They include *Helicobacter pylori*; *Streptococcus suis*; *Brucella suis*; *Campylobacter* spp., *Yersinia enterocolitica*, *Salmonella* spp.; *Listeria monocytogenes*, *Cryptosporidium parvum*, and *E. coli*. Another pathogen that has increased in prevalence in the developed world and is linked to swine farming is hepatitis E virus (HEV) (Clemente-Casares *et al.* 2003). Hepatitis E, an acute, self-limited, liver infection in humans, is a zoonotic disease (Smith 2001) with swine as the main animal reservoir. HEV, once endemic only in developing countries, has now been identified in the sewage of many industrialized countries including Spain, France and the U.S. (Kasorndorkbua 2004). There is also evidence showing that the human and swine strains of the virus are virtually identical (van der Poel *et al.* 2001; Banks *et al.* 2004), thus reinforcing the link between swine and human cases. Cole (2000) has also stated that commercial swine act as a reservoir for HEV and that human strains may have evolved from CAFO farms.

4.5.4 The Commercial/Industrial Sector

4.5.4.1 Power Generation

Water is a major requirement for producing thermoelectric power. Nuclear power plants continue to increase in number both in the U.S. and worldwide; and require between 500 million and 1.5 billion gallons (5.7 billion L) of water/day for proper operation (U.S. EPA 2003). Although water quality is not otherwise compromised, returned water is artificially heated causing thermal pollution (Parker 1979). According to Huizinga (1990) and Tyndall (1989), pathogenic strains of *Naegleria fowleri*, a free-living protozoan, have been isolated from both cooling towers, with subsequent human infections (Tyndal *et al.* 1983), as well as artificially-heated river waters (Figure 4.6). Although there have been no clearly documented cases of Naegleria or any other infection resulting from thermoelectric plant effluents, the U. S. Nuclear Regulatory Commission requires controls on the release of Naegleria spp. into the environment as it poses a potential health threat to those downstream. Recent analyses of the etiology of human cases of disease associated with recreational waters in the U.S. (Figure 4.3) reveal that *N. fowleri* accounted for 17% of outbreaks from untreated recreational waters (Craun *et al.* 2005).

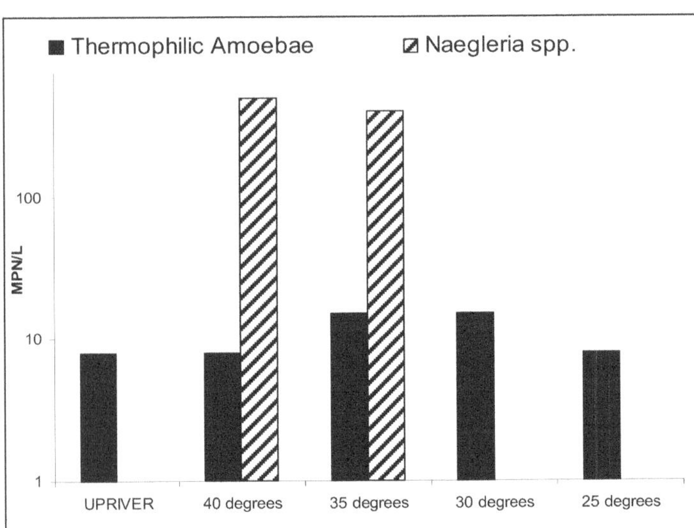

Figure 4.6 Presence of *Naegleria* spp. in Nuclear Power Plant Point Source Runoff (Tyndall *et al*, 1989); temperature in °C.

As is the case with thermoelectric power, the true impact of dams in general on waterborne spread of infections continues to be debated. For the majority of dams, the beneficiaries are large urban and industrial sectors which may be several to several hundred miles away. The impact, however, is usually felt locally in terms of environmental and societal change (WHO 2003) and the main pathogens of concern with dams are those spread by water-based vectors and secondary hosts (WHO 2003). In many developing countries, the spectrum of such diseases is quite wide, but in this region the prime focus is on the mosquito-borne infections such as West Nile fever (WNF); recently, a sharp rise in dead crows due to WNF was correlated with the presence of the Whittier Narrows Dam in the Los Angeles area (Reisen *et al.* 2004).

Although the chance of chemicals and microorganisms leaching from the basin of a dam into the groundwater is always a possibility, and evidence exists for heavy metal leaching from dam water to shallow groundwater (von der Heyden *et al.* 2004), there has been no record of pathogen spread in this manner.

4.5.4.2 Manufacturing and Mining

Unlike thermoelectric and hydroelectric plants, manufacturing and mining severely impact water through chemical and at times, microbial contamination. Water is used for various reasons but its major purpose is as a flushing agent, because when there is debris, a chemical contaminant or a biological nuisance that requires removal, water is the best option. This may result in chemical contamination of the watershed either through disposal or through leaching due to poor storage (Fent 2004). There is no data to directly associate infectious disease spread from water contaminated with manufacturing and mining discharges.

Effluents from pulp and paper mills have been shown to contain potential pathogens such as *Pseudomonas*, *Acinetobacter* and *Klebsiella* (Fulthorpe 1993; Gauthier *et al.* 2000), but such industrial discharges have not been associated with waterborne infections in humans (Gauthier and Archibald 2001). They may, however, have an indirect negative impact on humans due to a higher incidence of infections in fish species of economic importance (Khan 2004).

4.5.5 Climate Change

Irrespective of whether global warming is anthropogenic or not (Hansen *et al.* 1998), doubtless, it will significantly enhance the ability of water to spread infections by, among other things (a) increased intrusion of contaminated waters into otherwise clean freshwater resources, (b) widening the habitat of insect vectors and secondary hosts for human pathogens, (c) changes in microbial ecology of surface waters, increased algal growth, (d) unprecedented human population movements, and (e) shrinkage in accessible fresh water resources.

Rose *et al.* (1999) in the U.S. have assessed the influence of climate on the ability of water to spread infections and have defined a relationship between extreme precipitation and disease outbreaks from ground and surface waters. Such investigations identify precautions needed in vulnerable communities to preempt the risk of contamination of water sources and prevent outbreaks.

4.5.6 Other Factors

Eutrophication of receiving waters due to nutrient loadings and algal blooms is perhaps less common due to controls on sewage effluent discharges. However, and in contract to conventional wisdom, many algal species are now increasing being recognized as having direct and indirect impacts on human health (Hoeger *et al.* 2005; Peperzak 2005; Rehnstam-Holm and Hernroth 2005) and their potential in this context is expected to grow with global warming. Algae also harbor human enteric pathogens such as *Vibrio cholerae* (Thompson *et al.* 2004).

Drinking water distribution networks in urban settings in general provide a very efficient means for the dissemination of whatever may be in the water. This also makes such systems a potentially attractive means for rapid and widespread carriage of chemical, radiological and biological warfare agents (Meinhardt 2005). This is an emerging threat, but in reality though, the very nature and vastness of municipal water treatment and distribution systems also entail major impediments to their use for spreading bioweapons. Disinfectant residuals in the treated water may rapidly inactive many infectious agents introduced and, perhaps more importantly, the concentrations and volumes of such agents to achieve minimal infective doses in the water to be consumed are extremely difficult to attain.

4.6 CONCLUSION

Even though the assault on our water resources has been long-standing, the need for corrective action is more urgent now than ever before if we are to avoid an irreversible damage to the balance and quality of water. This review has attempted to objectively highlight mainly those issues which may be less familiar to other experts and policy makers in this region. Even though decisions on policy and action require a long-term vision, there are many instances where we, rushing to embrace a technology, either ignore or appear totally oblivious to its possible negative consequences. The experience with antibiotics, the proverbial 'magic bullets', is a case in point. That bacteria can develop resistance to antibiotics was known since the early days of the use of penicillin. Nonetheless, the essentially unchecked overuse and misuse of that and other antibiotics continued with the resultant and wide-spread antibiotic resistance. When it comes to water quality, the switch to chloramination from chlorination to reduce the risk of disinfection byproducts in municipal waters has likewise created other problems. As chloramine alters the microbial composition of biofilms in water distribution systems, a subsequent potential for consequences on human health results (Falkinham *et al.* 2003). Most likely, this change was introduced without any consideration of its impact on the microbiological quality of potable waters.

So, how can one prepare policy taking into account a potential that may or may not cause problems in the future? The answer is not as simple as taking a 'wait and see' approach. In the U.S. and Canada, conventional waterborne diseases such as cholera, typhoid and hepatitis A ceased to be a concern several decades ago. However, other types of infectious agents have filled this void, thus causing a change in the profile of waterborne pathogens (Theron and Cloete 2002). Good examples here are *C. parvum*, *T. gondii*, noroviruses, environmental mycobacteria and possibly hepatitis E virus. We believe that societal changes are in large part responsible for these emerging threats. We also continue to discover waterborne infectious agents that are refractory to conventional means of water treatment (Leclerc *et al.* 2002; Nwachcuku and Gerba 2004). In addition, and according to the data from WHO (2003), the etiology of many waterborne infections remains as yet undetermined (Figure 4.7). This points to the need for more effective and broad-spectrum approaches to water treatment rather than incremental improvements as and when we discover new problems. The combined and judicious application of UV irradiation, biological treatment, membrane filtration and reverse osmosis along with the existing and time-honored techniques of coagulation/filtration and disinfection should perhaps be considered (Goh 2005). Here again, it would be prudent to evaluate them for impacts such as those on biofilm quality prior to their widespread applications. Reducing the impact of biofilms on the microbiological quality of potable and recreational waters must constitute a priority area for research and development.

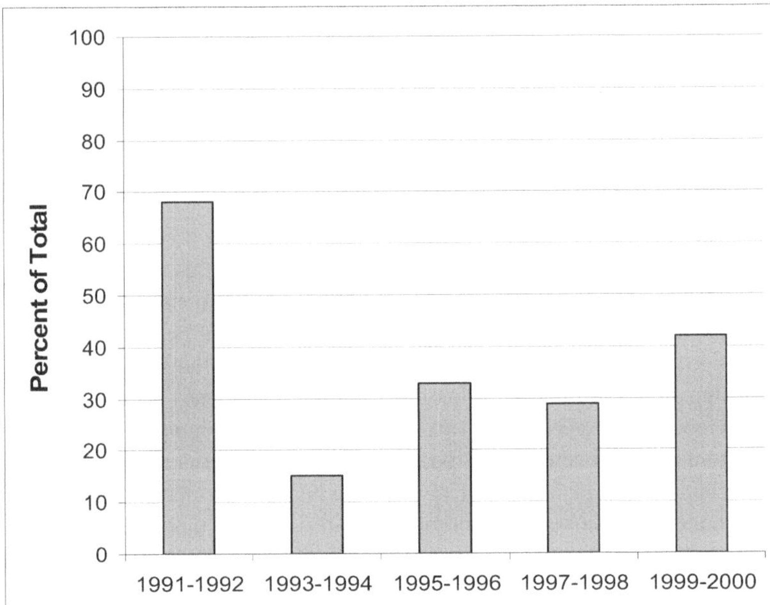

Figure 4.7. Percent of Global Waterborne Infections with Unknown etiology (World Health Organization, 2003)

The change in the profile of waterborne pathogens also questions the continued reliance on conventional indicators of the sanitary quality of potable waters (Rusin *et al.* 1997). Significant outbreaks of waterborne infections have been recorded in the two countries in spite of the fact the incriminated waters appeared to be free from fecal coliforms (Hrudy and Hrudy 2004).

Regional, national and international jurisdictions over marine and fresh water resources are behind the increasing numbers of cross-boundary disputes (de Villiers 1999; UNESCO 2003). The increasing demand on our shrinking fresh water resources is quite likely to enhance the frequency and severity of such disputes. Apart from its worldwide implications, this issue needs immediate attention in the U.S. and Canada also to develop better and more standardized means of collecting and disseminating water quality data and implementation of improvements and corrective measures with implications for the entire watershed than for local benefit only (Payment and Riley 2002).

Societal changes already in place and those which are bound to come will continue to influence our interactions with and exposure to microorganism (Woodward *et al.* 2000). If the immediate past is an indicator for things to come, it is highly likely that we will become more vulnerable to a variety to microbial pathogens. Since drinking water is an essential commodity, it is incumbent on us to ensure its safety. It is encouraging to note that the issue of water quality and its impact on human health is now receiving the attention at the highest level as enunciated in the following statement by Koïchiro Matsuura, the Director-General of UNESCO:

> *"Of all the social and natural crises we humans face, the water crisis is the one that lies at the heart of our survival and that of our planet Earth."*

4.8 REFERENCES

Ahmedna M, Marshall WE, Husseiny AA, Rao RM, Goktepe I. The use of nutshell carbons in drinking water filters for removal of trace metals. Water Res. 2004;38:1062-1068.

Anderson WB, Slawson RM, Mayfield CI. A review of drinking-water-associated endotoxins, including potential routes of human exposure. Can J Microbiol. 2002;48(7):567-587.

Angenent LT, Kelley ST, St Amand A, Pace NR, Hernandez MT. Molecular identification of potential pathogens in water and air of a hospital therapy pool. Proc Natl Acad Sci USA. 2005;102(13):4860-4865.

Aragon TJ, Novotny S, Enanoria W, Vugia DJ, Khalakdina A, Katz MH. Endemic cryptosporidiosis and exposure to municipal tap water in persons with acquired immunodeficiency syndrome (AIDS): a case-control study. BMC Public Health. 2003;3:2.

Aronson T, Holtzman A, Glover N, Boian M, Froman S, Berlin OG, Hill H, Stelma G Jr. Comparison of large restriction fragments of Mycobacterium avium isolates recovered from AIDS and non-AIDS patients with those of isolates from potable water. J Clin Microbiol. 1999;37:1008-1012.

Banks M, Bendall R, Grierson S, Heath G, Mitchell J, Dalton H. Human and porcine hepatitis E virus strains, United Kingdom. Emerg Infect Dis. 2004;10:953-955

Beuhler M. Potential impacts of global warming on water resources in southern California.Water Sci Technol. 2003;47:165-168.

Beuhler M. Tap water versus drinking water: is it time to have a debate? Water. 2002; 21:8-10.

Bloom MS, Vena JE, Swanson MK, Moysich KB, Olson JR. Profiles of ortho-polychlorinated biphenyl congeners, dichlorodiphenyldichloroethylene, hexachlorobenzene, and Mirex among male Lake Ontario sportfish consumers: the New York State Angler cohort study. Environ Res. 2005;97:178-194.

Bowie WR, King AS, Werker DH, Isaac-Renton JL, Bell A, Eng SB, Marion SA. Outbreak of toxoplasmosis associated with municipal drinking water. The BC Toxoplasma Investigation Team. Lancet. 1997;350(9072):173-7.

Canadian Food Inspection System. Code of hygienic practice for commercial prepackaged and non-prepackaged water. Ottawa, ON: Agriculture and Agri-Food Canada; 2003.

Castor ML, Beach MJ. Reducing illness transmission from disinfected recreational water venues: swimming, diarrhea and the emergence of a new public health concern. Pediatr Infect Dis J. 2004;23(9):866-870.

Chaidez C, Gerba C. Comparison of the microbiologic quality of point-of-use (POU)-treated water and tap water. Int J Environ Health Res. 2004;14:253-260.

Charron D, Thomas M, Waltner-Toews D, Aramini J, Edge T, Kent R, Maarouf A, Wilson J. Vulnerability of waterborne diseases to climate change in Canada: a review. J Toxicol Environ Health A. 2004;67:1667-1677.

Choi SM, Yoo SD, Lee BM. Toxicological characteristics of endocrine-disrupting chemicals: developmental toxicity, carcinogenicity, and mutagenicity. J Toxicol Environ Health B Crit Rev. 2004;7:1-24.

Clemente-Casares P, Pina S, Buti M, Jardi R, Martin M, Bofill-Mas S, Girones R. Hepatitis E virus epidemiology in industrialized countries. Emerg Infect Dis. 2003;9:448-454.

Cole D, Todd L, Wing S. Concentrated swine feeding operations and public health: a review of occupational and community health effects. Environ Health Perspect. 2000;108:685-699.

Coote DR, Gregorich LJ (eds.). The Health of Our Water: Towards sustainable agriculture in Canada. Ottawa, ON: Agriculture and Agri-Food Canada; 2000.

Corkal D, Schutzman WC, Hilliard C. Rural water safety from the source to the on-farm tap. J Toxicol Environ Health A. 2004;67(20-22):1619-1642.

Craun GF, Calderon RL, Craun MF. Outbreaks associated with recreational waters in the United States. International J Environ Hlth Res. 2005;15:243-262.

Crawford FG, Vermund SH. Human cryptosporidiosis. Crit Rev Microbiol. 1988;16:113-159.

De Villiers M. *Water*. Toronto: Stoddard; 1999.

Dubey JP. Toxoplasmosis — a waterborne zoonosis. Vet Parasitol. 2004 ;126:57-72.

Emmerson AM. Emerging waterborne infections in health-care settings. Emerg Infect Dis. 2001;7:272-276.

Environment Canada, 2003. http://www.nwri.ca/threatsfull/ch1-1-e.html.

Environment Canada. *Water Use*, 1996.
http://www.ec.gc.ca/water/en/manage/use/e_wuse.htm.

Ewald PW. *Evolution of Virulence*. Infect Dis Clin North Am. 2004;18:1-15.

Falkinham JO 3rd, Norton CD, LeChevallier MW. Factors influencing numbers of Mycobacterium avium, Mycobacterium intracellulare, and other Mycobacteria in drinking water distribution systems. Appl Environ Microbiol. 2001;67:1225-1231.

Falkinham JO 3rd. Mycobacterial aerosols and respiratory disease. Emerg Infect Dis. 2003;9:763-767.

Falkinham JO 3rd. Nontuberculous mycobacteria in the environment. Clin Chest Med. 2002;23:529-551.

Fent K. Ecotoxicological effects at contaminated sites. Toxicology. 2004 15;205:223-240.

Fleming P, Byers RH, Sweeney PA, *et al.* HIV prevalence in the United States, 2000 [Abstract 11]. Presented at the Ninth Conference on Retroviruses and Opportunistic Infections, Seattle, WA; February 24-28, 2002.

Ford-Jones EL, Wang E, Petric M, Corey P, Moineddin R, Fearon M. Rotavirus-associated diarrhea in outpatient settings and child care centers. The Greater Toronto Area/Peel Region PRESI Study Group. Pediatric Rotavirus Epidemiology Study for Immunization. Arch Pediatr Adolesc Med. 2000;154:586-593.

Fulthorpe RR, Liss SN, Allen DG. Characterization of bacteria isolated from a bleached Kraft pulp mill wastewater treatment system. Can J Microbiol. 1993;39:13-24.

Gale P. Land application of treated sewage sludge: quantifying pathogen risks from consumption of crops. J Appl Microbiol. 2005;98:380-396.

Gauthier F, Archibald F. The ecology of "fecal indicator" bacteria commonly found in pulp and paper mill water systems. Water Res. 2001;35:2207-2218.
Gauthier F, Neufeld JD, Driscoll BT, Archibald FS. Coliform bacteria and nitrogen fixation in pulp and paper mill effluent treatment systems. Appl Environ Microbiol. 2000;66:5155-51560.
Gea-Banacloche JC, Opal SM, Jorgensen J, Carcillo JA, Sepkowitz KA, Cordonnier C. Sepsis associated with immunosuppressive medications: an evidence-based review. Crit Care Med. 2004;32:S578-S590.
Gerba CP, Smith JE Jr. Sources of Pathogenic Microorganisms and Their Fate during Land Application of Wastes. J Environ Qual. 2005;34:42-48.
Germolec DR. Sensitivity and predictivity in immunotoxicity testing: immune endpoints and disease resistance. Toxicol Lett. 2004 Apr 1;149:109-114.
Ghannoum MA, O'Toole GA. *Microbial Biofilms*. 2004, ASM Press, Washington, DC.
Godfree A, Farrell J. Processes for managing pathogens. J Environ Qual. 2005;34:105-113.
Goedert JJ. Preventing infection-associated cancer: from bench to hillside. J Natl Cancer Inst. 2005;97:245-246.
Goh S, Reacher M, Casemore DP, Verlander NQ, Charlett A, Chalmers RM, Knowles M, Pennington A, William J, Osborn K, Richards S. Sporadic cryptosporidiosis decline after membrane filtration of public water supplies, England, 1996–2002. Emerg Infect Dis. 2005;11:251-259.
Guan TY, Holley RA. Pathogen survival in swine manure environments and transmission of human enteric illness—a review. J Environ Qual. 2003;32:383-392.
Hagedorn C, Robinson SL, Filtz JR, Grubbs SM, Angier TA, Reneau RB Jr. Determining sources of fecal pollution in a rural Virginia watershed with antibiotic resistance patterns in fecal streptococci. Appl Environ Microbiol. 1999;65:5522-5531.
Hansen JE, Sato M, Lacis A, Ruedy R, Tegen I, Matthews E. Climate forcings in the industrial era. Proc Natl Acad Sci U S A. 1998;95:12753-12758.
Health Canada. Canada's Aging Population. Division of Aging and Seniors. Health Canada; 2002.
Health Canada. HIV and AIDS in Canada. Surveillance report to December 31, 2002. Ottawa: Division of HIV/AIDS Epidemiology and Surveillance, Centre for Infectious Disease Prevention and Control, Health Canada, April 2003.
Health Canada. Report on the National Roundtable on Home and Community Care, February 10, 1999, Ottawa, ON: Health Canada; 1999.
Hoeger SJ, Hitzfeld BC, Dietrich DR. Occurrence and elimination of cyanobacterial toxins in drinking water treatment plants. Toxicol Appl Pharmacol. 2005;203(3):231-242.
Hrudey SE, Hrudey EJ. Safe Drinking Water: Lessons from Recent Outbreaks in Affluent Nations. London: IWA Publishing; 2004.
Hrudey SE, Payment P, Huck PM, Gillham RW, Hrudey EJ. A fatal waterborne disease epidemic in Walkerton, Ontario: comparison with other waterborne outbreaks in the developed world. Water Sci Technol. 2003;47:7-14.
Huizinga HW, McLaughlin GL. Thermal ecology of *Naegleria fowleri* from a power plant cooling reservoir. Appl Environ Microbiol. 1990;56:2200-2205.
Hunter PR. Drinking water and diarrhoeal disease due to *Escherichia coli*. J Water Health. 2003;1:65-72.

Hutin YJ, Pool V, Cramer EH, Nainan OV, Weth J, Williams IT, Goldstein ST, Gensheimer KF, Bell BP, Shapiro CN, Alter MJ, Margolis HS. A multistate, foodborne outbreak of hepatitis A. National Hepatitis A Investigation Team. N Engl J Med. 1999;340:595-602.

Hutson SS, Barber NL, Kenny JF, Linsey KS, Lumia DS, Maupin MA. Estimated Use of Water in the United States in 2000. USGS Circular 1268. Denver, CO: United States Geological Survey; 2004.

International Sanitary Supply Association. Report on 2002 Sanitary Supply Distributor Sales. Milwaukee: Trade Press Publishing Corporation; 2004.

Ismail A, Khosravi H, Olson H. R The role of infection in atherosclerosis and coronary artery disease: a new therapeutic target. Heart Dis. 1999;1:233-240.

Jasim S. Pharmaceutical and personal care products and endocrine disruptors — their occurrence, fate and transport in Great Lakes water supplies. Ontario Pipeline: Newsletter of the Ontario Water Works Assoc. 2004;8;10-11.

Johnson JY, Thomas JE, Graham TA, Townshend I, Byrne J, Selinger LB, Gannon VP. Prevalence of *Escherichia coli* O157:H7 and *Salmonella* spp. in surface waters of southern Alberta and its relation to manure sources. Can J Microbiol. 2003;49:326-335.

Juan CY, Thomas GO, Sweetman AJ, Jones KC. An input-output balance study for PCBs in humans. Environ Int. 2002;28:203-214.

Kasorndorkbua C, Guenette DK, Huang FF, Thomas PJ, Meng XJ, Halbur PG. Routes of transmission of swine hepatitis E virus in pigs. J Clin Microbiol. 2004;42:5047-5052.

Khan RA. Effect, distribution, and prevalence of Glugea stephani (Microspora) in winter flounder (Pleuronectes americanus) living near two pulp and paper mills in Newfoundland. J Parasitol. 2004;90:229-233.

Kirkhorn SR. Community and environmental health effects of concentrated animal feeding operations. Minn Med. 2002;85:38-43.

Kolankiewicz L, Beck R. 100 Largest U.S. Cities. Washington DC: SprawlCity.Org; 2000. http://www.sprawlcity.org/studyUSA/index.html.

Kummerer K. Drugs in the environment: emission of drugs, diagnostic aids and disinfectants into wastewater by hospitals in relation to other sources—a review. Chemosphere. 2001;45:957-969.

Lathers CM. Endocrine disruptors: a new scientific role for clinical pharmacologists? Impact on human health, wildlife, and the environment. J Clin Pharmacol. 2002;42:7-23.

Leblanc RJ, Allain CJ, Laughton PJ, Henry JG. Integrated, long term, sustainable, cost effective biosolids management at a large Canadian wastewater treatment facility. Water Sci Technol. 2004;49:155-162.

Leclerc H, Schwartzbrod L, Dei-Cas E. Microbial agents associated with waterborne diseases. Crit Rev Microbiol. 2002;28(4):371-409.

Lee SH, Levy DA, Craun GF, Beach MJ, Calderon RL. Surveillance for waterborne-disease outbreaks—United States, 1999-2000. MMWR Surveill Summ. 2002 22;51:1-47.

Levy CW, Roujeinikova A, Sedelnikova S, Baker PJ, Stuitje AR, Slabas AR. Molecular basis of triclosan activity. Nature. 1999;398:384-385.

Lewis DL, Gattie DK, Novak ME, Sanchez S, Pumphrey C. Interactions of pathogens and irritant chemicals in land-applied sewage sludges (biosolids). BMC Public Health. 2002;2:11.

MacKenzie WR, Hoxie NJ, Proctor ME, Gradus MS, Blair KA, Peterson DE, Kazmierczak JJ, Addiss DG, Fox KR, Rose JB. A massive outbreak in Milwaukee of cryptosporidium infection transmitted through the public water supply. N Engl J Med. 1994;331(3):161-167.

Macpherson CN. Human behaviour and the epidemiology of parasitic zoonoses. Int J Parasitol. 2005 Aug 12.

Mansfield LS, Gajadhar AA. Cyclospora cayetanensis, a food- and waterborne coccidian parasite. Vet Parasitol. 2004;126:73-90.

Marras TK, Daley CL.Epidemiology of human pulmonary infection with nontuberculous mycobacteria. Clin Chest Med. 2002;23(3):553-567.

Marras TK, Wallace RJ Jr, Koth LL, Stulbarg MS, Cowl CT, Daley CL. Hypersensitivity pneumonitis reaction to *Mycobacterium avium* in household water. Chest. 2005 Feb;127(2):664-71

McCaig LF, Besser RE, Hughes JM. Antimicrobial drug prescription in ambulatory care settings, United States, 1992-2000. Emerg Infect Dis. 2003;9:432-437.

Meinhardt PL. Water and bioterrorism: Preparing for the potential threat to U.S. water supplies and public health. Ann Rev Pub Hlth. 2005;26:213-237.

Merlani GM, Francioli P. R. Established and emerging waterborne nosocomial infections. Curr Opin Infect Dis. 2003;16:343-347.

Miao XS, Bishay F, Chen M, Metcalfe CD. Occurrence of antimicrobials in the final effluents of wastewater treatment plants in Canada. Environ Sci Technol. 2004;38:3533-3541.

Ministry of the Attorney General of Ontario. Part One Report of the Walkerton Commission of Inquiry. Toronto: Publications, ON; 2002a.

Ministry of the Attorney General of Ontario. Part Two Report of the Walkerton Commission of Inquiry. Toronto: Publications, ON; 2002b.

National Agricultural Statistics Service. 2003 Farm and Ranch Irrigation Survey. Washington, DC: NASS; 2004.

Niu MT, Polish LB, Robertson BH, Khanna BK, Woodruff BA, Shapiro CN, Miller MA, Smith JD, Gedrose JK, Alter MJ and Margolis, HS. Multistate outbreak of hepatitis A associated with frozen strawberries. J Infect Dis. 1992;166:518-524.

Norton CD, LeChevallier MW. A pilot study of bacteriological population changes through potable water treatment and distribution. Appl Environ Microbiol. 2000;66:268-276.

Nwachcuku N, Gerba CP. Emerging waterborne pathogens: can we kill them all? Curr Opin Biotechnol. 2004;15(3):175-180.

Nyquist AC, Gonzales R, Steiner JF, Sande MA. Antibiotic prescribing for children with colds, upper respiratory tract infections, and bronchitis. JAMA. 1998;79:875-877.

Parker FL. Thermal pollution consequences of the implementation of the president's energy message on increased coal utilization. Environ Health Perspect. 1979;33:303-314.

Patterson RA. Wastewater quality relationships with reuse options. Water Sci Technol. 2001;43:147-154

Payment P, Richardson L, Siemiatycki J, Dewar R, Edwardes M, Franco E. A randomized trial to evaluate the risk of gastrointestinal disease due to consumption of drinking water meeting current microbiological standards. Am J Public Health. 1991;81:703-708.

Payment P, Riley MS. Resolving the Global Burden of Gastrointestinal Illness: A Call to Action. Washington, DC: American Academy of Microbiology; 2002.

Peperzak L. Future increase in harmful algal blooms in the North Sea due to climate change. Water Sci Technol. 2005;51(5):31-36.

Pieniazek NJ, Bornay-Llinares FJ, Slemenda SB, da Silva AJ, Moura IN, Arrowood MJ, Ditrich O, Addiss DG. New cryptosporidium genotypes in HIV-infected persons. Emerg Infect Dis. 1999;5:444-449.

Primm TD, Lucero CA, Falkinham JO. Health impacts of environmental mycobacteria. Clin Microbiol Rev. 2004;17:98-106.

Pryor M, Springthorpe S, Riffard S, Brooks T, Huo Y, Davis G, Sattar SA. Investigation of opportunistic pathogens in municipal drinking water under different supply and treatment regimes. Water Sci Technol. 2004;50:83-90.

Rehnstam-Holm AS, Hernroth B. Shellfish and public health: a Swedish perspective. Ambio. 2005;34(2):139-144.

Reisen W, Lothrop H, Chiles R, Madon M, Cossen C, Woods L, Husted S, Kramer V, Edman J. West Nile virus in California. Emerg Infect Dis. 2004;10:1369-1378.

Reiss R, Mackay N, Habig C, Griffin J. An ecological risk assessment for triclosan in lotic systems following discharge from wastewater treatment plants in the United States. Environ Toxicol Chem. 2002;21:2483-2492.

Rose CS, Martyny JW, Newman LS, Milton DK, King TE Jr, Beebe JL, McCammon JB, Hoffman RE, Kreiss K. "Lifeguard lung": endemic granulomatous pneumonitis in an indoor swimming pool. Am J Public Health. 1998;88(12):1795-1800.

Rose JB, Daeschner S, Easterling DR, Curriero FC, Lele S, Patz JA. Climate and waterborne disease outbreaks. J Am Water Works Assoc. 1999;92:77-87.

Rose JB, Epstein PR, Lipp EK, Sherman BH, Bernard SM, Patz JA. Climate variability and change in the United States: potential impacts on water- and foodborne diseases caused by microbiologic agents. Environ Health Perspect. 2001;109 Suppl 2:211-221.

Rosenblum LS, Mirkin IR, Allen DT, Safford S, Hadler SC. A multifocal outbreak of hepatitis A traced to commercially distributed lettuce. Am J Public Health. 1990;80:1075-1079.

Rusin PA, Rose JB, Haas CN, Gerba CP. Risk assessment of opportunistic bacterial pathogens in drinking water. Rev Environ Contam Toxicol. 1997;152:57-83.

Russell AD. Mechanisms of antimicrobial action of antiseptics and disinfectants: an increasingly important area of investigation. J Antimicrob Chemother. 2002;49:597-599.

Sargeant M. Challenges, Risks, and Rewards: Learning to Control Our Biological Fate. In Biomedicine and the Human Condition Challenges, Risks and Rewards. Cambridge: Cambridge University Press; 2005:1-19.

Saskatchewan Justice. The North Battleford Water Inquiry. Regina, SK: Govt. Printing, 2002.

Sattar SA, Tetro J, Springthorpe VS. Impact of changing societal trends on the spread of infections in American and Canadian homes. Am J Infect Control. 1999;27:S4-S21.

Schneider E, Hajjeh RA, Spiegel RA, Jibson RW, Harp EL, Marshall GA, Gunn RA, McNeil MM, Pinner RW, Baron RC, Burger RC, Hutwagner LC, Crump C, Kaufman L, Reef SE, Feldman GM, Pappagianis D, Werner SB. A coccidioidomycosis outbreak following the Northridge, Calif, earthquake. JAMA. 1997;277:904-908.

Schneider K, Hafner C, Jager I. Mutagenicity of textile dye products. J Appl Toxicol. 2004;24:83-91.

Schwab BW, Hayes EP, Fiori JM, Mastrocco FJ, Roden NM, Cragin D, Meyerhoff RD,

D'Aco VJ, Anderson PD. Human pharmaceuticals in US surface waters: a human health risk assessment. Regul Toxicol Pharmacol. 2005;42:296-312.

September SM, Brozel VS, Venter SN. Diversity of nontuberculoid Mycobacterium species in biofilms of urban and semi-urban drinking water distribution systems. Appl Environ Microbiol. 2004;70:7571-7573.

Shiklomanov, IA. World Water Resources and their Use: A Joint SHI/UNESCO Product. St. Petersburg: State Hydrological Institute (SHI); 1999.

Simmons J, Bourne LS. The Canadian Urban System, 1971-2001: Responses to a Changing World. Toronto: Centre for Urban and Community Studies, University of Toronto; 2003.

Sischo WM, Atwill ER, Lanyon LE, George J. Cryptosporidia on dairy farms and the role these farms may have in contaminating surface water supplies in the northeastern United States. Prev Vet Med. 2000;43:253-267.

Slifko TR, Smith HV, Rose JB. Emerging parasite zoonoses associated with water and food. Int J Parasitol. 2000;30: 1379-1393.

Smith HV, Rose JB. Waterborne cryptosporidiosis. Parasitol Today. 1990;6:8-12.

Smith JL. A review of hepatitis E virus. J Food Prot. 2001;64:572-586.

Smolinski MS, Hamburg MA, Lederberg J. *Microbial Threats to Health: Emergence, Detection, and Response*. Institute of Medicine. 2003. The National Academies Press, Washington, DC.

Sobsey MD Field survey of enteric viruses in solid waste landfill leachates. Am J Public Health. 1978;68:858-864.

Stagnitti MN. Statistical Brief #21 Trends in Outpatient Prescription Drug Utilization and Expenditures: 1997-2000. Washington: Medical Expenditure Panel Survey; 2003.

Stanley K, Jones K. Cattle and sheep farms as reservoirs of Campylobacter. J Appl Microbiol. 2003;94 Suppl:104S-113S.

Statistics Canada. Guide to health statistics. Ottawa, Canada: Statistics Canada; 2000.

Steenland K, Bertazzi P, Baccarelli A, Kogevinas M. Dioxin revisited: developments since the 1997 IARC classification of dioxin as a human carcinogen. Environ Health Perspect. 2004;112:1265-1268.

Stevenson KB, Loeb M. Performance improvement in the long-term-care setting: building on the foundation of infection control. Infect Control Hosp Epidemiol. 2004;25:72-79

Theron J, Cloete TE. Emerging waterborne infections: contributing factors, agents, and detection tools. Crit Rev Microbiol. 2002;28:1-26.

Thompson FL, Iida T, Swings J. Biodiversity of vibrios. Microbiol Mol Biol Rev. 2004;68(3):403-431,

Tyndall RL, Ironside KS, Metler PL, Tan EL, Hazen TC, Fliermans CB. Effect of thermal additions on the density and distribution of thermophilic amoebae and pathogenic Naegleria fowleri in a newly created cooling lake. Appl Environ Microbiol. 1989;55:722-732.

Tyndall, R. L. Chlorination as an Effective Treatment for Controlling Pathogenic Naegleria in Cooling Waters of an Electric Power Plant. Water Chlorination 1983;4(2):1097-1103.

United Nations Educational, Scientific and Cultural Organization (UNESCO). Water for People, Water for Life. New York, NY: The United Nations; 2003.

United Nations Environment Programme. *The Global Mercury Assessment Report.* 2003. http://www.chem.unep.ch/mercury/Report/Final%20Assessment%20report.htm.

United States Census Bureau, 2004. U.S. Interim Projections by Age, Sex, Race, and Hispanic Origin. http://www.census.gov/ipc/www/usinterimproj.

United States Census Bureau. Who's Minding the Kids? Child Care Arrangements: Spring 1999. Washington: Fertility and Family Statistics Branch; 2003.

United States Centers for Disease Control and Prevention. 2002. http://www.cdc.gov/mmwr/preview/mmwrhtml/ss5108a1.htm.

United States Environmental Protection Agency Office of Research and Development. Research Plan for Endocrine Disruptors. Washington, DC: EPA; 1998.

United States Environmental Protection Agency. National Pollutant Discharge Elimination System Permit Regulation and Effluent Limitation Guidelines and Standards for Concentrated Animal Feeding Operations (CAFOs). Fed Reg. 2003;68:7175-7274.

United States Food and Drug Administration. Part 165: Beverages. Fed Reg. 2003;21:521-539.

van der Poel WH, Verschoor F, van der Heide R, Herrera MI, Vivo A, Kooreman M, de Roda Husman AM. Hepatitis E virus sequences in swine related to sequences in humans, The Netherlands. Emerg Infect Dis. 2001;7:970-976.

van Leeuwen FXR. Safe drinking water: the toxicologist's approach. Food Chem Toxicol. 2000;38:S51-S58.

Vial T, Nicolas B, Descotes J. Clinical immunotoxicity of pesticides. J Toxicol Environ Health. 1996;48:215-229.

von der Heyden CJ, New MG. Groundwater pollution on the Zambian Copperbelt: deciphering the source and the risk. Sci Total Environ. 2004;327:17-30.

Wagner D, Young LS. Nontuberculous mycobacterial infections: a clinical review. Infection. 2004;32:257-270.

Wallace R, Wallace D. Origins of public health collapse in New York City: the dynamics of planned shrinkage, contagious urban decay and social disintegration. Bull N Y Acad Med. 1990;66(5):391-434/Council on Environmental Quality. The Urban River, In Environmental Quality Along the American River, Department of the Environment, 1996.

Wallace RJ Jr, Brown BA, Griffith DE. Nosocomial outbreaks/pseudo-outbreaks caused by nontuberculous mycobacteria. Annu Rev Microbiol. 1998;52:453-490.

Weiss RA, McMichael AJ. Social and environmental risk factors in the emergence of infectious diseases. Nat Med. 2004;10:S70-76.

Wilson S, Dzon L, Reed A, Pruitt M, Whalen MM. Effects of in vitro exposure to low levels of organotin and carbamate pesticides on human natural killer cell cytotoxic function. Environ Toxicol. 2004;19:554-563.

Woodward A, Hales S, Litidamu N, Philips D, Martin J. Protecting human health in a changing world: the role of social and economic development. Bull WHO. 2002;78:1148-1155.

World Health Organization Emerging Issues in Water and Infectious Disease. Geneva: WHO; 2003.

World Health Organization. Human health and dams: the World Health Organization's submission to the World Commission on Dams (WCD). Geneva: WHO Press; 2000.

World Health Organization. The world health report 2002 - Reducing Risks, Promoting Healthy Life. Geneva: World Health Organization; 2002.

Yang S, Carlson K. Evolution of antibiotic occurrence in a river through pristine, urban and agricultural landscapes. Water Res. 2003;37:4645-4656.

5

Managing Human Impacts on Water Quality: Role of Integrated Ocean Observing Systems

T. C. Malone and D. Rockwell

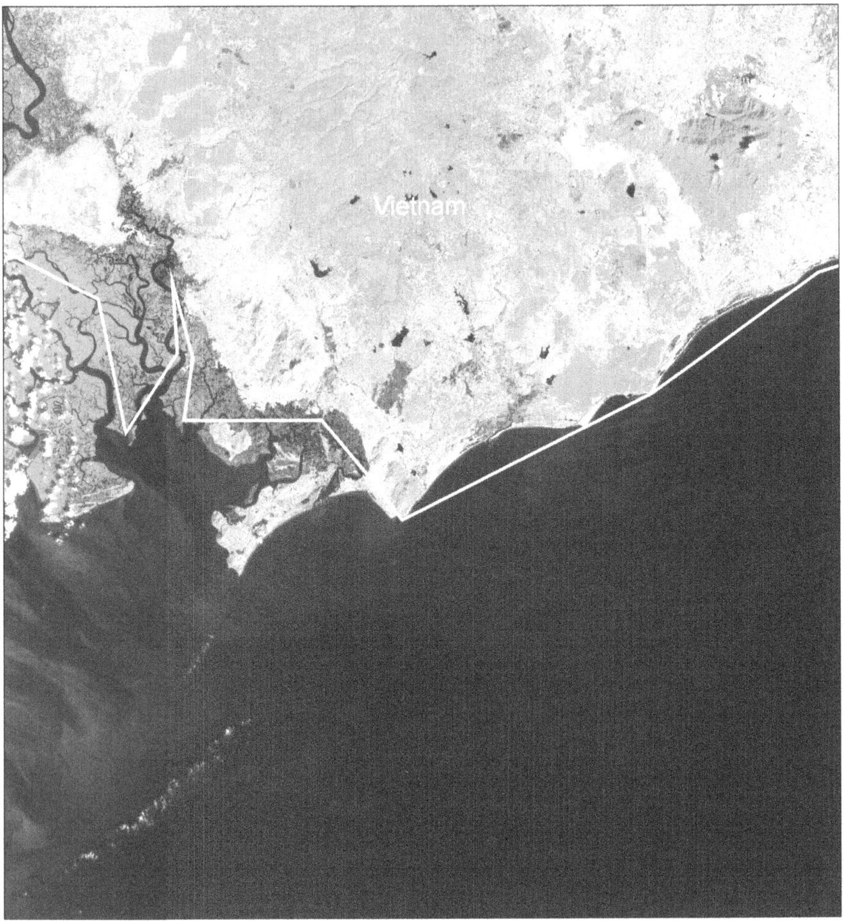

Mekong River. Image: MSU Center for Global Change and Earth Observations

5.1 SUMMARY

The risks of human exposure to waterborne pathogens are greatest in coastal waters where the effects of natural hazards, human activities and climate change converge. Managing and reducing risks depend on rapid detection and timely prediction of changes in the distribution and abundance of waterborne pathogens, harmful algal species, and their toxins. The provision of and access to data and information required for rapid detection and timely prediction is a goal of the coastal module of the Global Ocean Observing System. Case studies of monitoring systems for harmful algal blooms in the Gulf of Mexico and waterborne pathogens in the Great Lakes underscore current inadequacies including the lack of (1) real-time *in situ* sensing capabilities for biological (abundance of pathogens) and chemical (concentration of toxins) properties; (2) data assimilating, coupled physical-ecological numerical models for nowcasting and forecasting distributions of pathogens and toxins; and (3) rapid access to relevant data from both *in situ* measurements and remote sensing. These inadequacies highlight the need for integrated and sustained observations of coastal waters; integrated data management and communications; and for research and development that will produce the next generation of *in situ* sensors and ecosystem models needed for more effective decision making to reduce risks of exposure and increase the economic value of beaches and coastal ecosystems worldwide.

5.2 INTRODUCTION

This contribution is concerned primarily with the coastal module of the Global Ocean Observing System (GOOS) and its potential for improving assessments of risks of human exposure to waterborne pathogens and harmful algal blooms. Section 5.3 gives the rationale for establishing GOOS; section 5.4 summarizes the conceptual design of GOOS; and section 5.5 uses case studies to illustrate requirements for the observing system and potential applications made possible by GOOS development. The concluding section highlights priorities for research and development needed to meet these requirements and achieve the systems potential.

5.3 HEALTH RISKS IN AN ECOSYSTEM CONTEXT

5.3.1 Climate, People and Coastal Ecosystems

The cumulative effects of natural hazards, human activities, and climate change are and will continue to be most pronounced in the coastal zone (Throughout this chapter, "coastal zone" and "coastal waters" include marine and estuarine systems and the Great Lakes.) where people and ecosystem goods and services are most concentrated, exposure to natural hazards is greatest, and inputs of energy and matter from land, sea and air converge (Costanza *et al.* 1993; McKay and Mulvaney 2001; Nicholls and Small 2002; Small and Nicholls 2003). The changes occurring in coastal waters ("phenomena of interest") affect public health and wellbeing, the safety and efficiency of marine operations, and the capacity of ecosystems to support goods and services including the sustainability of living marine resources and biodiversity (Table 5.1). Although these phenomena tend to be local in scale, they are occurring in coastal ecosystems worldwide and are often local expressions of larger scale variability and change including both natural and anthropogenic forcings:

- Natural hazards and extreme weather (Michaels, *et al.* 1997; Epstein 1999; Flather 2000);
- Global warming and sea level rise (Barry *et al.* 1995; Najjar *et al.* 1999; Levitus *et al.* 2000);
- Basin scale changes in ocean-atmosphere interactions (El Niño Southern Oscillation, North Atlantic Oscillation and Pacific Decadal Oscillation) (Barber and Chavez 1986; Wilkinson *et al.* 1999; Arcos *et al.* 2001; Koblinsky and Smith 2001; *Lipp et al.* 2001; Beaugrand *et al.* 2002; Francis *et al.* 1998);
- Human alterations of the environment (Peierls *et al.* 1991; Vitousek *et al.* 1997; GESAMP 2001; Heinz Center 2002),
- Exploitation of living resources (Jackson *et al.* 2001; Myers and Worm, 2003), and
- Introductions of non-native species (Carlton 1996; Hallegraeff 1998).

Each of these drivers of change has been shown to influence human health risks from exposure to waterborne pathogens to the toxins produced by harmful algae (HA).

The clearest and most direct impacts of these drivers on the oceans and human health occur in coastal areas that are subject to intense human use (sewage discharge, agriculture and aquaculture, human habitation) and are susceptible to flooding from tsunamis, storm surges, and excessively heavy rainfall associated with tropical storms and monsoons (NRC 1999). Thus, the global increase in population density in coastal areas (Nicholls and Small 2002; Small and Nicholls 2003) is correlated with a global increase in the incidence of human diseases in coastal populations due to a combination of factors ranging from increasing discharges of raw sewage and ballast water from ships (that introduce pathogens and harmful algae) to extreme weather and associated flooding and storm water runoff (NRC 1999). There is also increasing evidence that global scale changes in the abundance and distribution of both waterborne and vector-borne diseases are occurring in response to global warming and associated changes in rainfall patterns (Haines and Parry 1993; Rogers and Packer 1993; Colwell 1996; and Epstein 1999).

Table 5.1 Examples of the drivers of change (forcings) and associated "phenomena of interest" in coastal waters of the oceans and Great Lakes. Forcings and phenomena of interest that are related to or may increase the probability of human exposure to waterborne pathogens and harmful algal toxins are underlined.

	FORCINGS OF INTEREST
"Natural"	Global warming, sea level rise Natural hazards (tropical storms (cyclones), tsunamis) Currents, waves, tides and storm surges River and groundwater discharges, sediment inputs
Anthropogenic	Alteration of hydrological and nutrient cycles Inputs of chemical contaminants and human pathogens Harvesting natural resources (living and nonliving) Physical alterations of the environment Introductions of non-native species
	PHENOMENA OF INTEREST
Climate and weather	Variations in sea surface temperature; surface fluxes of momentum, heat and fresh water; sources and sinks of carbon; sea ice
Marine operations	Variations in water level, bathymetry, surface winds, currents and waves; sea ice; susceptibility to natural hazards
Natural hazards	Storm surge and coastal flooding; coastal erosion; susceptibility to natural hazards; public safety and property loss
Public health	Risk of exposure to human pathogens, chemical contaminants, and biotoxins (contact with water, aerosols, seafood consumption)
Healthy Ecosystems	Habitat modification, loss of biodiversity, cultural eutrophication, harmful algal events, invasive species, chemical contamination, diseases in and mass mortalities of marine organisms
Living marine resources	Fluctuations in spawning stock size, recruitment and natural mortality; changes in areal extent and condition of essential habitat; food availability Aquaculture production and water quality

Cholera epidemics illustrate how changes in ocean-atmosphere interactions on the scale of the large ocean basins can affect human health risks through events such as the El Niño-Southern Oscillation (ENSO) and global warming (Colwell 1996; NRC 1999). Outbreaks of cholera have been traced to the discharge of ballast water from ships, and annual epidemics of cholera in Bangladesh have been related to high water temperatures, water level, turbidity and copepod abundance. *Vibrio cholerae* is a waterborne pathogen that inhabits the guts of copepods. Because pathogens and nutrients are transported into coastal waters during high run off events, ENSO-related increases in rainfall and high runoff in the aftermath of coastal flooding caused by tsunamis and tropical storms are likely to promote plankton blooms and increase the risk of human exposure to pathogens and toxins produced by harmful algae. In the case of cholera, remote sensing of sea surface temperature, chlorophyll-a concentration and sea surface height can be used to map exposure risk (NRC 1999). In terms of human alterations, Jackson *et al.* (2001) and Lenihan *et al.* (2001) recently argued that over-fishing is not only the first major human perturbation of coastal ecosystem dynamics, it exacerbates the effects of nutrient pollution, contributes to habitat loss and associated loss of biodiversity, and increases the susceptibility of coastal ecosystems to invasive species, harmful algal blooms (HABs) and outbreaks of disease.

5.3.2 Managing Human Use in the Context of Natural Variability and Change

Successful prevention, management and mitigation of the environmental effects of human activities, natural hazards and climate change depend on the capacity to anticipate changes with sufficient lead-time to make informed decisions with desired outcomes (Clark *et al.* 2001). To this end, ecosystem-based management, which considers the effects of human activities in the context of natural variability, is emerging as a unifying approach to environmental protection, resource management, land-use planning and environmental engineering (NRC 1999; Sherman and Duda 1999; UNEP 2001; Cicin-Sain and Knecth 1998). Implementing such a strategy requires the capability to engage in *adaptive management*, a decision making process that depends on reliable, routine and rapid detection of changes in the condition or state of coastal ecosystems. As indicated by recent attempts to quantify the condition of coastal marine and estuarine ecosystems on regional to global scales, *we do not have this capability today* (Malone *et al.* 2005).

Ecosystem-based approaches are especially important in coastal waters where the combined effects of habitat alterations, land-based inputs of nutrients and contaminants, over fishing, harmful algal blooms, and invasive species are most severe (Botsford *et al.* 1997). Effective ecosystem-based management of the risks of exposure to waterborne pathogens and HABs depends on rapid detection and timely predictions. Rapid assessment requires improved observing capabilities that increase the speed and accuracy with which changes in the distributions of key environmental parameters, pathogens and HA species are detected and assessed. Timely predictions require both improved data assimilating models of ecosystem dynamics and observations to parameterize and initialize them (Figure 5.1).

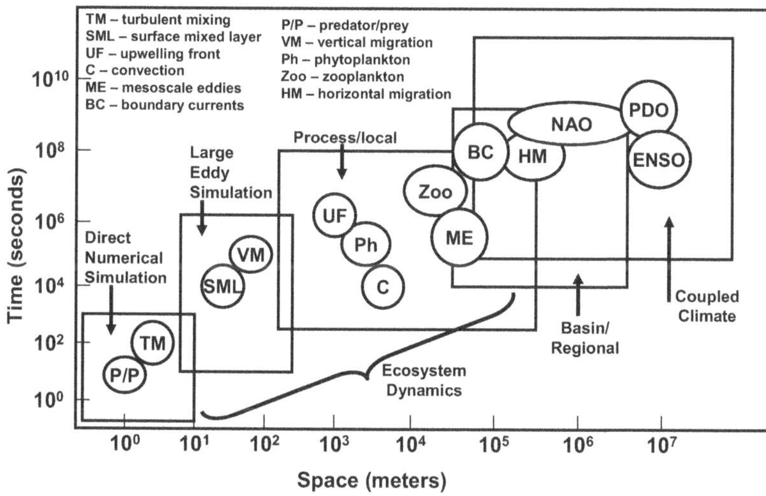

Figure 5.1. Time-space scales of marine processes and models that link global climate dynamics on large scales to the population dynamics of phytoplankton on the smaller scales of biomass accumulation and trophic interactions within ecosystems. The rectangles approximate the time-space scales encompassed by global ocean-climate and basin scale models (coupled climate, basin/regional) and smaller scale models of ecosystem dynamics ranging from the mesoscale (e.g., tidal and upwelling fronts and associated phytoplankton and zooplankton aggregations) to the scales of turbulent mixing and predator-prey interactions (modified from Montua et al. 2002).

5.4 THE GLOBAL OCEAN OBSERVING SYSTEM

Emergence of the science of ecological forecasting (Clark *et al.* 2001; Hofmann and Friedrichs 2002) and operational oceanography, in the form of the Global Ocean Observing System (GOOS), are important steps toward developing the capability to engage in ecosystem-based management practices. The mandate to establish a Global Ocean Observing System was articulated and ratified as an international consensus in 1992 with the signing of the Framework Convention on Climate Change and the Program of Action for Sustainable Development at the U.N. Conference on Environment and Development (UNCED) (IOC 1998; http://ioc.unesco.org/goos/Prospe98/Contents.html; http://ioc.unesco.org/goos/docs/GOOS_125_COOP_Plan.pdf).

The latter calls for the establishment of a GOOS that will enable effective management and sustainable use of the marine environment and its natural resources. GOOS is a major component of the Integrated Global Observing Strategy (IGOS) and the oceans and coasts component of the Global Earth Observing System of Systems (GEOSS) (IOC 1998; IOC 2003; www.earthobservationsummit.gov/declaration.html). In December, 2004, the President of the United States issued an Executive Order for the establishment of a Cabinet level Commission on Ocean Policy and an Integrated Ocean Observing System (IOOS) as the U.S. contribution to GOOS (http://www.whitehouse.gov/news/releases/2004/12/20041217-5.html). The Ocean Action Plan prepared in response to recommendation of the U.S. Commission on Ocean Policy can be found at http://ocean.ceq.gov/actionplan.pdf.

5.4.1 Purpose

GOOS is developing as a global network that systematically and continuously acquires and disseminates data and information based on requirements specified by groups that use, depend on, manage or study marine and estuarine systems (IOC 1998; IOC 2003). The purpose of GOOS is to make possible more effective use of data and information from existing monitoring programs (by providing more rapid access to data from many different sources, including data from both *in situ* measurements and remote sensing; enabling the development of operational ecosystem models; etc.) and to enhance these efforts as needed by decision makers. To these ends, GOOS is being developed in two interdependent modules:

1. an open ocean module concerned primarily with detecting and predicting natural hazards and changes in the ocean-climate system and their effects on marine operations and services (http://ioc.unesco.org/goos/docs/GOOS_066_act.pl.htm), and
2. a coastal module concerned primarily with effects of human activities, natural hazards and climate change on public health risks, the condition of coastal marine and estuarine ecosystems, and the living resources they support (IOC 1998; IOC 2003).

GOOS is being designed and implemented to routinely and continuously provide data and information needed to achieve six related societal goals:

- Improve regional forecasts of the weather and predictions of warming, precipitation, and sea level patterns associated with global climate change;
- Improve the safety and efficiency of marine operations;
- Mitigate the effects of extreme weather (e.g., tropical storms) more effectively;
- Minimize public health risks;
- Protect and restore healthy ecosystems more effectively; and
- Sustain living marine resources.

Achieving these goals requires rapid detection and timely prediction of a broad spectrum of phenomena (Table 5.1) each of which have unique requirements for data and information. At the same time, many requirements for observations and data management are similar across all six goals. Thus, although the focus of this contribution in on public health risks associated with waterborne pathogens and harmful algae, it should become clear that an integrated approach to the development of a multi-use, multi-disciplinary observing system is feasible, sensible and cost-effective.

5.4.2 An End-to-End System

Rapid detection and timely prediction depend on the establishment of an observing system that efficiently links measurements and modeling via integrated data communications and management. The system must be sustained (permanent) to provide continuity and to capture the spectra of variability that characterize the marine environment and the organisms that inhabit it. It must be integrated to provide spatially and temporally synoptic observations of physical, chemical and biological variables to achieve multiple objectives encompassed by the six goals. Linking user needs to measurements requires an efficient and managed flow of data and information among three essential subsystems (Figure 5.2):

1. A monitoring subsystem (coastal networks for near shore observations and land-based remote sensing; the global network of coastal tide gauges; ships, fixed platforms, moorings, drifters, and underwater vehicles for *in situ* measurements; remote sensing from satellites and aircraft) to measure variables on time-space scales required to assess the state of marine ecosystems and to rapidly detect and predict changes in or the occurrence of priority phenomena of interest (Table 5.1);
2. A data communications network and management subsystem that provides rapid access to diverse data from many sources;

3. A data-assimilation and analysis (conceptual, statistical and numerical models; geographic information systems) subsystem for the provision of data and information in formats and at rates specified by the users.

Together, the subsystems constitute an "end-to-end" system that provides data and information for informed decisions on time scales required to achieve desired affects and to evolve in response to the changing needs of multiple user groups.

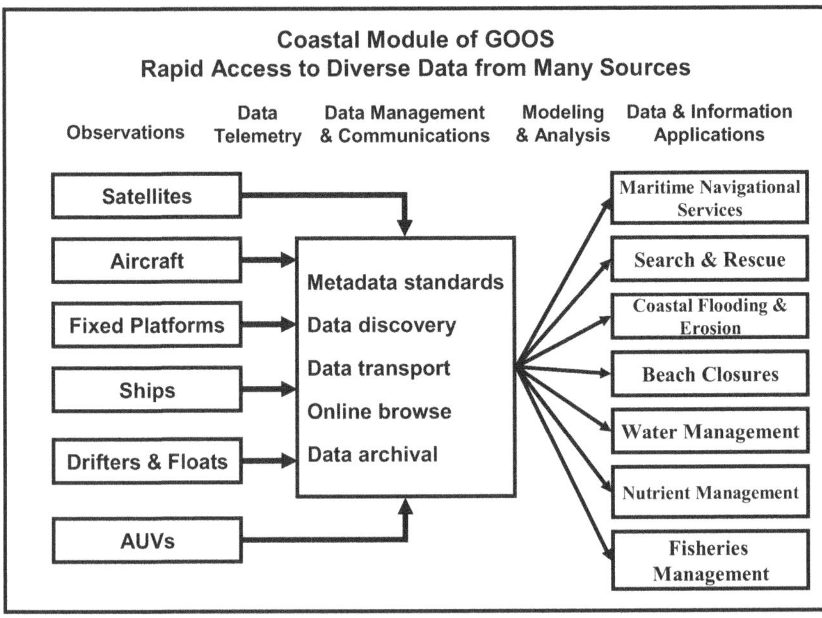

Figure 5.2. The global atmospheric observing system of the World Weather Watch (WWW) is the kind of operational, end-to-end system that GOOS is envisioned to be. Just as weather nowcasts and forecasts are not possible without sustained observations and operational models, implementation of ecosystem-based, adaptive management practices will not be possible without operational models of marine ecosystems and uninterrupted provision of oceanographic data required to initialize and up-date model runs. However, the analogy ends here as GOOS will serve a broad diversity of multidisciplinary data and information and its development will require a broader base of scientific disciplines. (AUV – Autonomous Underwater Vehicle)

5.4.3 The Coastal Component of the Global Ocean Observing System

The Integrated Design Plan for the Coastal Module of GOOS (IOC 1998; IOC 2003) calls for establishing regional coastal ocean observing systems (RCOOSs) worldwide (Figure 5.3) and, through this process, the development of a Global Coastal Network (GCN). The GCN will (1) measure, manage and analyze common variables needed by all or most coastal nations and regions; (2) establish a sparse network of sentinel and reference stations for early detection of the effects of land-based inputs (e.g., freshwater, sediments and nutrients) and basin scale variability (e.g., the El Niño-Southern Oscillation, North Atlantic Oscillation, and Pacific Decadal Oscillation); and (3) apply internationally accepted standards and protocols for measurements, data telemetry, data management and modeling.

The provisional common variables include geophysical variables (temperature, salinity, currents, waves, sea level, shoreline position, bathymetry, sediment grain size), chemical variables (dissolved inorganic nutrients, dissolved oxygen, sediment organic content), biological variables (fecal indicators, phytoplankton biomass, benthic biomass), and biophysical variables (optical properties). Depending on national and regional priorities, GRAs may increase the resolution at which the common variables are measured, supplement common variables with the measurement of additional variables and provide data and information products that are tailored to the requirements of stakeholders in the respective regions. Although fecal indicators are to be measured by the GCN, it is likely that sampling programs that augment the GCN and target specific fecal indicators will be developed by participating nations as part of regional coastal ocean observing systems. Likewise, because HA species differ from region to region, it is expected that the GCN will be similarly augmented for priority HA species in each region.

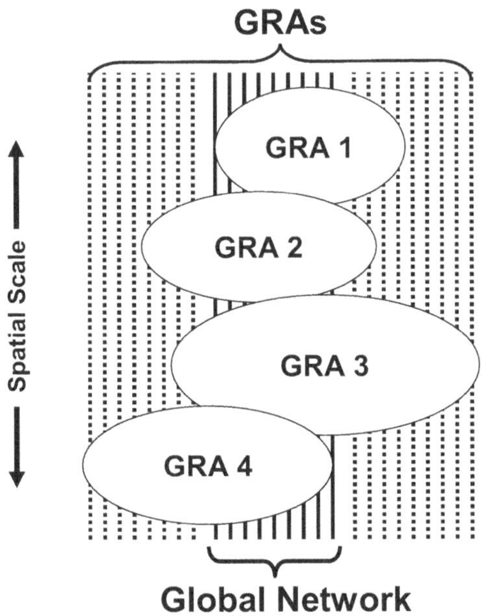

Figure 5.3. (A) Schematic relationship between the global network of coastal observations and GOOS Regional Alliances (GRAs) established to oversee the design, implementation, operation and development of regional observing systems. Solid vertical lines represent the common variables to be measured and processed as part of the global network. Dashed lines represent regional enhancements by GRAs (discs), such as sea ice in polar regions, or coral characteristics in the tropics. Geographic boundaries of regional systems are not fixed in space; they may overlap; and they are determined by the time and space scales of the phenomena of interest that are high priorities in each region.

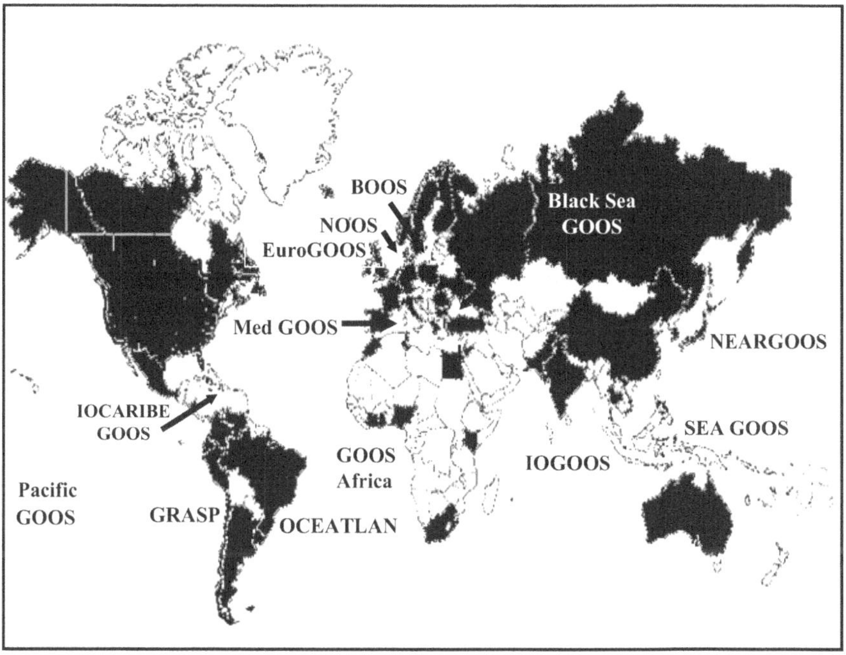

Figure 5.3 (B) GOOS Regional Alliances (black) and countries with national GOOS Programmes (black) are forming worldwide. This map shows their location as of 2003.

5.5 ECOSYSTEM-BASED APPROACHES TO MANAGING HEALTH RISKS

The oceans and Great Lakes are conduits for many pathogenic microorganisms and their toxins (Table 5.2). Their distributions and exposure risks in aquatic systems are governed by their sources; their behavior in aquatic media (from freshwater to sea water) once introduced (rates of growth, mortality, migration, buoyancy, etc.); their place in the food web; and by water motions that transport, disperse or concentrate them. The most effective ways to reduce the immediate cost of lives and human suffering from exposure to waterborne pathogens and HABs is to control sources (e.g., reduce inputs of untreated sewage wastes, reduce land-based inputs of anthropogenic nutrients, reduce the temporal and spatial extent of coastal flooding), detect increases in risk and associated environmental conditions more rapidly, and provide timely accurate predictions of increases in risk in both time and space.

Increases in risk to levels that lead to beach and shellfish bed closures are typically localized, episodic and dynamic. Consequently, rapid, timely, and accurate assessments of risk are difficult if not impossible based on traditional sampling regimes, e.g., monthly or bi-weekly monitoring of sewage outfalls and daily shoreline sampling at a limited number of beach sites. Remote sensing and the development of species-specific *in situ* sensors for waterborne pathogens and HABs have great potential for providing the means to address these challenges. For example, satellite-based Synthetic Aperture Radar (SAR) provides high resolution (< 100 m) active microwave observations of sea surface roughness that are independent of cloud cover and time of day. At surface wind speeds between 2 m sec^{-1} and 7 m sec^{-1}, areas with biogenic or anthropogenic surfactant films are detected by SAR as patterns of low backscatter return because such films dampen small waves. Recent studies using the Southern California Bight as a model system illustrate the ability of SAR to detect and track the fate of stormwater runoff and sewage discharge (Svejkovsky and Jones 2001; DiGiacomo *et al.* 2004). In combination with field surveys, land-based high frequency radar, and numerical models, these studies demonstrate the potential for rapid detection and timely predictions that can be used to inform management and mitigation decisions that reduce public health risks and increase the economic and social value of beaches and living resources.

Two case studies are described below, one for HABs and one for fecal indicators of waterborne allochthonous pathogens. Both illustrate the common need for hydrodynamic models (and meteorological and physical oceanographic observations) and the challenges and potential of developing early warning observing systems for more cost-effective management and mitigation of their impacts (e.g., reducing health risks and realizing economic benefits through fewer closures of shellfish beds and beaches).

Table 5.2. Selected lists of agents of waterborne disease, their pathologies, and routes of transmission (modified from NRC 1999).

Agent	Pathology	Transmission Route
PATHOGENS		
Viruses		
Hepatitis A	Infectious hepatitis	Seafood[1], Water ingestion
Hepatitis B	Hepatitis	Water ingestion
Caliciviruses		
Rotaviruses	Gastroenteritis	Water ingestion
Astroviruses	Gastroenteritis	Seafood[1], Water ingestion
Enteroviruses	Gastroenteritis	Water ingestion
Autochthonous Bacteria		
Mycobacterium marinum	Granuloma	Water contact
Vibrio alginolyticus	Wound infections	Water contact
V. cholerae	Cholera	Seafood[1], Water ingestion
V. parahaemolyticus	Gastroenteritis, Wound infections	Seafood[1], Water contact

Table 5.2. Selected lists of agents of waterborne disease, their pathologies, and routes of transmission (modified from NRC 1999).

Agent	**Pathology**	**Transmission Route**
Allochthonous Bacteria		
Escherichia coli	Dysentery, Gastroenteritis	Water contact
Leptospira interrogans	Leptospirosis	Water contact
Listeria monocytogenes	Listeriosis	Seafood[1]
Morganella morganii	Scomboid food poisoning	Seafood[1]
Salmonella spp.	Typhoid, Gastroenteritis	Water ingestion
Shigella spp.	Dysentery	Water ingestion
HARMFUL ALGAE		
Dinoflagellates		
Alexandrium catenella	Paralytic shellfish poisoning	Seafood[2]
A. tamaense	Paralytic shellfish poisoning	Seafood[2]
Gymnodinium catenatum	Paralytic shellfish poisoning	Seafood[2]
Pyrodinium bahamense	Paralytic shellfish poisoning	Seafood[2]
Karenia brevis	Neurotoxic shellfish poisoning	Seafood[2, 3]
Gambierdiscus toxicus	Ciguatera fish poisoning	Seafood[2]
Prorocentrum spp.	Ciguatera fish poisoning	Seafood[2]
Prorocentrum spp.	Diarrheic shellfish poisoning	Seafood[2]
Dinophysis spp.	Diarrheic shellfish poisoning	Seafood[2]
Diatoms		
Pseudo-nitzschia multiseries	Amnesic shellfish poisoning	Seafood[2]
P. australis	Amnesic shellfish poisoning	Seafood[2]
Cyanobacteria		
Anabena spp.	Diarrhea	Water ingestion
Microcystis spp.	Diarrhea	Water ingestion
Oscillatoria spp.	Diarrhea	Water ingestion

[1] Consumption of raw or undercooked seafood.

[2] Cooked or uncooked.

[3] Inhalation of aerosol particles coated with the toxin can also cause respiratory problems (tightness of breath, coughing and sneezing, tearing eyes, burning sensation in mucous membranes and excess mucous discharge.

5.5.1 Harmful Algal Bloom Observing Systems

5.5.1.1 Gulf of Mexico

Concerns about harmful algal blooms (HABs) have increased over the last decade largely because of the perceived increase in the number and duration of events. The toxins produced by these species cause finfish and shellfish poisoning, a variety of human pathologies that can lead to death, and mass mortalities of marine organisms including fish, mammals and birds. Recently, the economic impacts of HAB outbreaks along Chinese coast have been estimated to be > $1.2 0 billion US. A HAB (NSP) outbreak in 1987-88 closed more than 400 km of North Carolina coastline for shellfishing during the peak harvesting season, causing economic losses estimated at $25 million.

In the Gulf of Mexico, the red tide species *Karenia brevis* routinely occurs along the southwest coast of Florida in the late summer and early fall and can persist for up to three months (Steidinger *et al.* 1997). More sporadic blooms occur along the coast from Texas to North Carolina. Aquatic organisms are affected either by the neurotoxin itself or the by the reduced water quality that results from a bloom. In 1996, a bloom was responsible for the death of 10% of the manatee population. Human health is compromised by the presence of dead and decaying fish in the waters and on the beach and by the production of aerosols that cause asthma-like symptoms. The blooms impact fisheries and tourist industries by inducing neurotoxic shellfish poisoning. In the 2002-2003 HAB season, the Florida shellfish aquaculture and oyster industries lost $6 million in dockside sales alone, and up to 20% of the planted clams. More rapid detection and timely predictions of HABs will reduce economic losses and the risk of human exposure.

The case study described here is the Gulf of Mexico Harmful Algal BloomS Observing System (HABSOS) Project launched in 2000 following a 1997 workshop on harmful algal bloom forecasting and a 2000 workshop on linking management requirements to the development of GOOS (Fisher *et al.* 2003; Stumpf *et al.* 2003; http://ccma.nos.noaa.gov/rsd/projects/habs/html, www.csc.noaa.gov/crs/habf). The goal of HABSOS is to create an internet-based data communications and management system for collecting, processing and disseminating data and information for more rapid detection (early alerts, frequent updates of the locations of existing blooms) and timely predictions of HABs (forecasts of bloom trajectories, time and location of land fall, and probabilities of where and when a bloom will occur based on environmental conditions).

HABSOS is designed and implemented as a proof of concept, user-driven project. To this end, the project is a partnership involving U.S. federal and state agencies, academic marine research laboratories in the U.S. and Mexico, and industries working in the Gulf of Mexico. Current primary users are government agencies responsible for public health officials, management of living marine resources and environmental protection.

The Gulf of Mexico has many features that make this region suitable for a pilot project that can rapidly develop into an operational system if successful. These include the following:

- There is a long history of monitoring HABs (Figure 5.4) and associated shellfish bed closures;
- Although home to 30 or more toxic species, *Karenia brevis* (formerly *Gymnodinium breve*) has by far the greatest impact in the region in terms of human health, mass mortalities of marine organisms, and shellfish bed closures (Steidinger *et al.* 1997);
- *K. brevis* can be detected by both remote and autonomous *in situ* sensing (Schofield *et al.* 1999; Hu *et al.* 2003; Stumpf *et al.* 2003). It is pigmented with concentrations of chlorophyll *a* and cells correlated over the dynamic range of concentrations typically found in the Gulf (< 1 – 20 μg Chl liter^{-1}); it has a surface expression that can be detected by satellite-based remote sensing; and it has a specific signature of inherent optical properties that allows recognition by *in situ* optical instrumentation (Stumpf *et al.* 2003);
- *K. brevis* blooms are a regional problem, (e.g., they are not confined to one state but are often transported across state jurisdictions) (Tester *et al.* 1991; Stumpf *et al.* 2003);
- *K. brevis* blooms occur most frequently during late summer and fall when diatom biomass tends to be low and *K. brevis* often dominates the chlorophyll *a* field (Kusek *et al.* 1999);
- Blooms of *Trichodesmium* spp., a genus of cyanobacteria with an optical signature (phycobilins) different from that of *K. brevis* (Subramaniam and Carpenter, 1994), often precede *K. brevis* blooms (Walsh and Steidinger 2001); and
- The frequency and timing of blooms may be a useful indicator of global warming (Tester *et al.* 1993).

A comprehensive review of the occurrence of *K. brevis* blooms, the biology and ecology of *K. brevis* and its impacts (real and perceived) is given by Kusek *et al.* (1999).

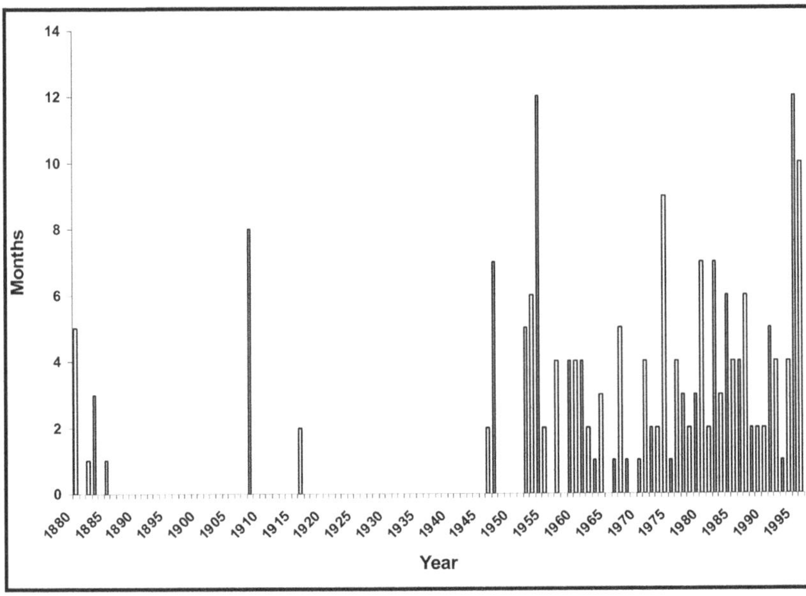

Figure 5.4. Long-term time series of red tides (duration of *Karenia brevis blooms* year^{-1}) off the west coast of Florida, USA. About 70% of blooms lasted between 1 and 4 months with 15% lasting between 8 and 12 months. No secular trend is apparent.

5.5.1.2 Detection

A climatological approach to detecting chlorophyll *a* anomalies that are most likely caused by *K. brevis* (75% of the time during 1999 through 2001) has been developed by Stumpf (2001) and used to provide early warnings of new blooms, to track the movements of existing blooms, and to forecast trajectories (Stumpf *et al.* 2003). An anomaly is identified when differences between the current chlorophyll field (from SeaWiFS imagery obtained in real time from OrbImage) and a mean field exceeds 1 μg liter-1 (equivalent to 100 cells ml^{-1}), where the mean is a two month running mean ending two weeks prior to the current image. Two months provides a sufficient number of images over a long enough period to be representative of the season. A 2-week lag minimizes the possibility that a persistent and stationary bloom will bias the mean. This information is communicated to user groups in near real-time via a web-based HAB Bulletin.

5.5.1.3 Prediction

The ultimate goal is to predict the probability of a bloom in time and space and when and where it is likely to impact people and fisheries. This requires knowledge of the organism's physiology and environmental conditions that are likely to lead to and transport blooms. *K. brevis* has been reported to be both positively and negatively phototactic (Steidinger 1975; Kamykowski *et al.* 1998). Blooms along the west coast of Florida often appear during upwelling-favorable winds when preceded by a period of calm winds (Stumpf *et al.* 1998). Although it is believed that blooms are initiated offshore, blooms often appear along the shoreline with no indication of offshore development based on satellite imagery (Stumpf *et al.* 2003). Together, these observations suggest that blooms are initiated and develop offshore at depth under calm conditions. With the development of wind-driven upwelling circulation, blooms are transported onshore with bottom water and transported to the surface by coastal upwelling. Based on this hypothesis an alert was issued in the HAB Bulletin and a nearshore bloom was successfully forecasted in September, 2000 (Stumpf *et al.* 2003).

Developing capabilities such as this are an important step toward the development of models that predict where and when a bloom is likely to be initiated based on environmental conditions. The minimum set of observations that will be needed to achieve this objective are as follows:

1. forcings: real-time wind fields and precipitation updated 3 to 4 times daily; river and stream flows and associated transports of sediments and nutrients updated weekly.
2. ecosystem properties: surface current, temperature, salinity and chlorophyll fields updated daily; vertical profiles of temperature, salinity, dissolved oxygen, inorganic nutrients, chlorophyll, colored dissolved organic matter, and *K. brevis* cell densities at weekly (hot spots that have a history of frequent HABs) to monthly intervals (larger area, more stations).

The elements required to develop this capacity to provide the required data streams, data communications and management, and models exist in the form of the NOAA HAB remote sensing team (http://ccma.nos.noaa.gov/rsd/projects /habs/html, www.csc.noaa.gov/crs/habf), Texas Coastal Ocean Observation Network (TCOON) and the Texas Automated Buoy System (TABS) operated by Texas AandM University (http://dnr.cbi.tamucc.edu/, http://gergu3 .tamu.edu/Tglo/), meteorological buoys and stations maintained by the National Data Buoy Center (http://seaboard.ndbc.noaa.gov/), the Coastal Ocean Monitoring and Prediction System (COMPS) of the University of South Florida (http://comps.marine.usf.edu/), the Sustained Ecological Research Related to Management of the Florida Keys (SEAKEYS) (http://coral.aoml. noaa.gov/sferpm/seakeys/), the National Estuarine Research Reserves (NERRS) (http://inlet.geol.sc.edu/cdmohome.html), and a data management initiative of the NOAA National Coastal Data Development Center (www.ncddc.noaa.

gov/habsos). Collaboraton among these efforts is a priority of the emerging Gulf Coast Ocean Observing System (http://www-ocean.tamu.edu/GCOOS/).

5.5.2 Fresh and Brackish Water Ecosystems

Harmful algal blooms are not limited to marine waters. Recent years have seen an expansion of toxic blooms in fresh-brackish water ecosystems, e.g., the Great Lakes (Erie and Ontario) and the Baltic Sea. Since 1995, toxic cyanobacteria of the genus *Microcystis* have bloomed annually in Lake Erie's western basin and produced potent hepatotoxins (microcystins) that exceed the threshold advisory concentration established by the World Health Organization. The Great Lakes contain almost 10% of the world's freshwater and serve as a source of drinking water for an estimated 22 million people. Protecting this and other freshwater resources are essential. Because microcystins are extremely potent, the U.S. Environmental Protection Agency includes cyanobacteria and cyanobacterial toxins on their "Contaminant Candidates List" which requires the development of analytical methods for measuring their concentration and solicitation of data needed to assess the occurrence of these toxins in drinking water and the health effects of exposure to them. As has been demonstrated in the Baltic Sea (http://www.fimr.fi/en/itamerikanta/bsds/819.html), a combination of *in situ* measurements (using ferries and other ships of opportunity as platforms) and satellite-based remote sensing of ocean color could become the basis for a Great Lakes HABSOS much like the Gulf of Mexico HABSOS.

5.5.3 Beach Closures in the Great Lakes

The global burden of human disease caused by sewage pollution of coastal waters is estimated at 4 million lost 'man-years' annually. This is roughly equivalent to an annual economic loss of approximately \$16 billion U.S. (http://www.gpa.unep.org/pollute/sewage.htm). The annual income from recreational activities in the world's coastal zone (including the use of bathing beaches and seafood consumption) is estimated to be on the order of \$100 billion U.S. annually. On a local scale, net economic losses due to beach closures have been estimated to range from \$1,274 to \$37,030 per day for one Lake Michigan beach depending value assumptions used (Rabinovici *et al.* 2004). Improved estimates of land-based inputs and models of coastal circulation and water quality will reduce risks of human exposure, provide the data and information needed for more effective control of anthropogenic inputs, and maximize recreational income (Cheves 2003).

5.5.3.1 The Decision Making Process

Closures of beaches due to risks of human exposure to pathogens has been a priority issue for the Great Lakes National Program Office (GLNPO) since 1981 (Jacobs 1997) and is one of eight critical environmental issues that has received national attention (Executive Order 13340). Human health and recreational use of beaches can be adversely affected by water borne pathogens in the Great Lake Ecosystem.

Land-based sources of pathogens and bacterial indicators of pathogens include both point source discharges of untreated waste water (e.g., combined sewer overflows, SSO, boat sanitary discharges, and septic systems) and non-point (diffuse) surface runoff of fecal contaminants from wildlife and live stock. Both inputs are exacerbated by excessive rainfall (e.g., associated with tropical storms and ENSO events) and coastal inundation (e.g., storm surge flooding and tsunamis). Once introduced into aquatic systems, determination of health risks and subsequent beach notifications or swim closures depends on timely and accurate estimates of concentrations of waterborne pathogens, their distribution, and changes in distribution over time.

Because of uncertainties associated with both estimates of pathogen concentrations and distributions, public officials face the probability that the decision to close a beach will be in error since little correlation is found between measurements from one day to next day and distributions change with time due to circulation, mixing, and mortality (Olyphant *et al.* 2003). The waters at the beach may be safe for swimming, but the official closes the waters; or the waters at the beach may be unsafe for swimming and kept open in error. Since the official is likely to use the precautionary principle, more swimming at beaches is likely to be prevented when they could be open if uncertainty were reduced.

5.5.3.2 Detection

Because of the high diversity of waterborne microbial pathogens and the cost of direct measurements of their concentration, *Escherichia coli* (a fecal indicator) is used in the United States to indicate the risk of exposure to pathogens in freshwaters (Dufour 1984). Current cost effective tests using culture techniques require 18-24 hours to determine concentrations of *E. coli*. Thus, decisions to close a beach or take no action (keep the beach open and permit swimming) are based information that does not become available until 18-24 hours after the fact. Errors are also introduced because concentrations of fecal indicators such as *E. coli* are often not correlated with concentrations of microbial pathogens themselves (NRC, 1999). This is exacerbated by the fact that some strains of *E. coli* produce human toxins and can survive in a viable but unculturable state for days to weeks (NRC, 1999).

5.5.3.3 Prediction

Pathogens of human origin (e.g., allochthonous bacteria, Table 5.2) enter coastal waters with surface runoff and sewage discharges. Once in coastal waters, changes in the distribution and concentration of human pathogens is determined primarily by the "half life" of pathogens and their rates of transport, concentration, and dispersion. Pathogens introduced via storm water runoff and point source discharges are typically concentrated in buoyant plumes that can be observed and modeled to provide nowcasts of location and forecasts of transport and dilution. As discussed above, an optimal mix of *in situ* observations and remote sensing can be used to provide the required observations. Predictive models have been developed to estimate the dilution, dispersion and decline of fecal indicators (IOC 2003). These coupled physical-biological models incorporate the effects of (1) dilution due to turbulent mixing, (2) advective transport, and (3) the rate of decay in toxicity or mortality (including the dormancy rate).

In the Great Lakes, direct measurement of surface winds and currents have been used in an experimental mode to provide nowcasts and forecasts of *E. coli* distributions that can be used by beach managers to advise the swimming public more effectively. Model predictions of *E. coli* distribution use wind speed and direction, wave height, water turbidity, suspended sediment concentration, rainfall in the preceding 24 hours, and water temperature to make such forecasts (Francy and Darner 1998; Francy *et al.* 2003; Whitman *et al.* 1999). Great Lake Beaches have numerous and varying conditions that affect bacterial conditions (Nevers and Whitman 2004), but a regional nowcast model for southern Lake Michigan is being developed that shows clusters of beaches responding in similar ways. Monitoring systems provide daily metrological measurements that are assimilated into a nowcast model with prediction accuracy ranging from 54-91% depending on beach zone and high or low predictions (Whitman 2005). The data and information provided by an integrated and sustained Great Lakes observing system coupled with beach specific models will reduce these uncertainties leading to fewer beach closures and lower health risks.

5.5.4 Toward More Informed Decisions Through Rapid Detection and Timely Prediction

Current approaches for managing public use of beaches and near-shore waters for swimming suffer from three sources of uncertainty: (1) the time required to detect *E. coli* indicators of pathogens (24 – 48 hours); (2) the need to measure indicator concentrations in the context of environmental parameters that determine their distribution and concentration; and (3) need for operational, data assimilating models that can provide nowcasts and forecasts of the distribution of pathogens using coupled hydrodynamic-biological models. Reducing these sources of uncertainty requires (1) the development of species-specific, inexpensive methods of rapidly measuring the concentration of waterborne pathogens (rapid detection); and (2) observations and models of the aquatic environment that can be used to nowcast and forecast their distributions (timely prediction).

To address these problems, the U.S. EPA, USGS and NOAA are collaborating to support an integrated effort to develop an observing system for more timely and accurate forecasts of the distribution and abundance of waterborne pathogens in beach environments. An important part of this effort is the development of a Beach Forecasting Tool (BFT) (Frick *et al.* 2004). BFT combines data on bacteria sources and concentrations with hydrodynamic and statistical models to generate 3-day forecasts of beach concentrations. In this interagency collaboration, NOAA provides a wind-driven, 3-dimensional hydrodynamic model (Beletsky *et al.* 2003); EPA provides point source plume models (Frick *et al.* 2003); and EPA and USGS have compiled data and developed source strength and mortality models to enable initial bacterial concentrations and decay rates to be estimated (Frick *et al.* 2003; Olyphant *et al.* 2003; and Myers *et al.* 1998).

BFT is the basis of "Visual Beach" which will make forecasts available via the world wide web and will disseminate forecasts to registered users (Frick *et al.* 2004). Monitoring results will be used to initialize model runs and to verify model results. The goal is to overcome the long time lag between changes in environmental concentrations of *E. coli* and their detection (24-48 hours), help guide adaptive sampling, and to inform the public and decision makers responsible for beach closures. The effectiveness of this approach is illustrated during high and low flows of the Milwaukee River in May and June, 2004 (Figure 5.5 A and B). The integration of three dimensional hydrodynamic models using wind direction modeling, wind direction and speed measurements, and geosynchronous orbiting satellites can provide the means to predict and verify the growth, movement and dissipation of buoyant plumes and possible their potential impact on coastal beaches. Linkage of these hydrodynamic models, forecast and now cast models to computer web sites offer the promise of providing monitoring and modeling results so the swimming community can plan their weekend. (Frick *et al.* 2004).

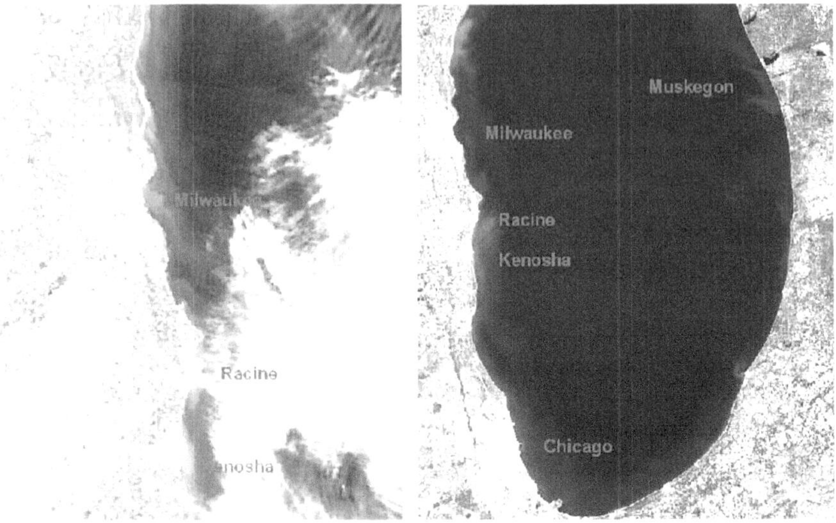

Figure 5.5. (A) On 25 May 2004, the sediment plume of the Milwaukee River is visible in the LandSat image on the left. By 3 June, 2004 the plume has dissipated (Frick *et al.* 2004).

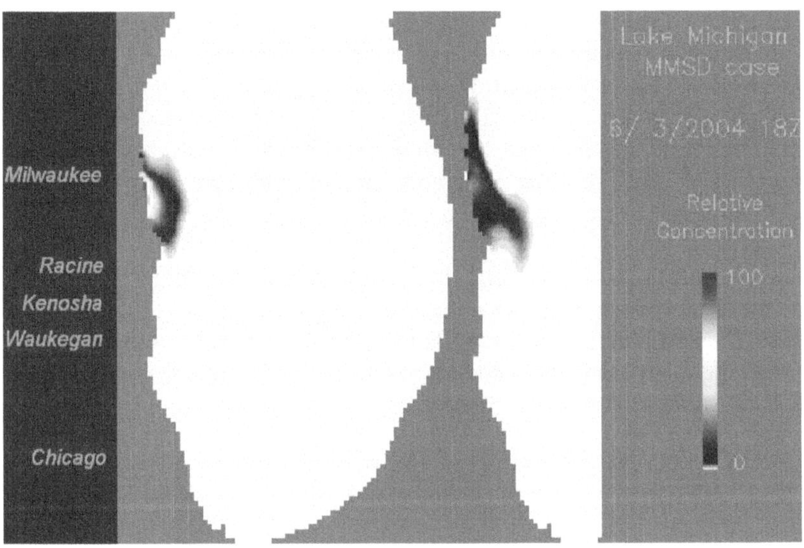

Figure 5.5. (B) Plume simulations using the NOAA hydrodynamic model for May (left) and June (right) correspond well with observations shown in Figure 5a (Frick *et al.* 2004).

5.6 CONCLUSIONS

Substantial progress has been made in the design and implementation of the open ocean module of GOOS over the last 15 years (Koblinsky and Smith, 2001).[2, 5] In contrast, although a high priority of the international community, development of the coastal module has been slow.[2] This is due in part to (1) the lack of capacity to rapidly and routinely detect changes in those phenomena of interest that require measurements of biological and chemical variables and (2) the lack of operational ecosystem-based models required to predict the occurrence of or changes in ecological phenomena of interest (Figure 5.6).

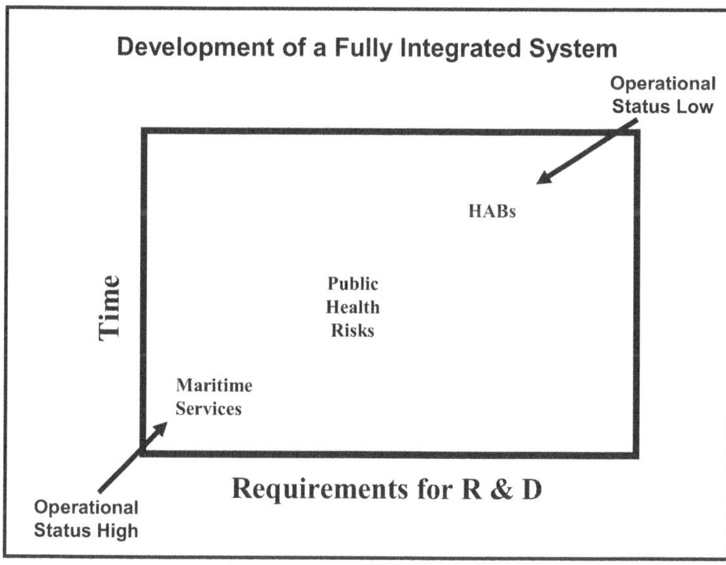

Figure 5.6. Time-dependent development of a fully integrated observing system. Predictions of most of the phenomena of interest (Table 1) require the same physical and meteorological data acquired for marine services and forecasting natural hazards. Those elements of an operational observing system required for improved marine services and forecasts of natural hazards are most developed (including the required operational models) while those required for ecosystem-based management of public health risks, ecosystem health (as indicated by the frequency and magnitude of HABs in this case), and living resources are least developed. Capabilities to assess public health risks and the status of marine ecosystems are somewhere in between. Given current capabilities and the importance of physical processes to the phenomena of interest relevant to the goals of ecosystem health and living marine resources, initial development of the global system will target the measurement and processing of physical variables <u>and</u> those biological and chemical variables that can be measured routinely and for which products can be clearly defined. To the extent that they are operational, non-physical variables should be incorporated into the observing system now (e.g., chlorophyll fluorescence, dissolved

oxygen, fecal indicators). As technologies and procedures for rapid measurement and analysis of additional biological (e.g., phytoplankton species, zooplankton abundance, fish abundance, concentrations of pathogens, biomarkers) and chemical properties (e.g., nutrients, pesticides, PAHs) are developed, they will be incorporated into the system to achieve regional priorities and meet the needs of user groups.

Improving the operational capabilities of GOOS depends on major advances on at least three research fronts: (1) development of integrated data communications and management systems that provide rapid access to diverse data from many sources; (2) development of *in situ* and remote sensing techniques for near real-time provision of data on key biological and chemical variables; and (3) development of data assimilation techniques and coupled physical-ecological models that can be run in an operational mode. Harmful algal blooms and waterborne pathogens are important cases in point.

An important goal of research on and monitoring of waterborne pathogens and HABs is to provide the knowledge, data and information needed to protect public health and sustain the capacity of marine ecosystems to support goods and services. Ecosystem-based management strategies are aimed at preventing HABs (e.g., reduce nutrient loading) and pathogen contamination (e.g., sewage treatment) before they occur, mitigating their effects (e.g., close shellfish beds, close beaches, move net pens of cultured salmon), and controlling them once they occur (reducing their magnitude, containing their distribution). Achieving these objectives requires the development of three related capabilities for both implementing adaptive management and assessing their efficacy:

1. more rapid detection of waterborne pathogens, HA species and toxins,
2. timely predictions of where and when public health risks are likely to be unacceptably high, and
3. timely forecasts of trajectories and contaminated water masses in time and space.

Rapid detection is a high priority for both waterborne pathogens and HABs and molecular techniques offer a way forward. For example, species-specific DNA probes from ribosomal sequences have the potential to provide accurate and rapid diagnostic tools for the evaluation of environmental samples (NRC, 1999). When combined with polymerase chain reaction (PCR), these probes allow detection of an increasing number of pathogens and indicators. The recent application of "real-time" PCR to field diagnostics of microbial pathogens reveals the potential of this approach for rapid and reliable detection of pathogens in aquatic systems.

To the extent that allochthonous waterborne pathogens behave as passive particles with known half-lives and their sources are well documented and quantified, the development of operational nowcasting and forecasting capabilities depends primarily on more rapid detection of pathogens and increases in the spatial resolution of hydrodynamic models. The HAB challenge is more complex. It will not be possible to develop operational models for HAB prediction based on environmental conditions until the combination of environmental factors that promote the growth and accumulation of one species over others are quantified and physical-biological interactions are parameterized (e.g., how environmental factors such as turbulence, advective transport, light, nutrient, and grazing, and inherent biological attributes interact with each other to favor the development of a given species). Developing this capacity will require significant advances in our understanding of the processes of species succession; in the development of coupled, data assimilating physical-ecological models; and in the capacity to estimate the distribution and abundance of HA species and toxins rapidly and synoptically with time-space estimates of physical and chemical fields using a combination of *in situ* and remote sensing techniques. These include (1) more accurate estimates of sea surface chlorophyll *a* and accessory pigment fields on space scales of < 1 km based on ocean leaving radiance measurements from satellites and aircraft (improve the skill of case 2 algorithms and increase spatial resolution); (2) long-term, high resolution time series by instrumenting moorings and fixed platforms with sensors to measure apparent optical properties and nutrient concentrations (N, P, Si) synoptically with temperature, salinity, currents and waves; (3) techniques for rapid, species-specific identification and enumeration, including near real-time measurement and telemetry of HA cell densities; (4) techniques for more rapid measurement of HA toxins, including *in situ* detection and near real-time telemetry and (5) rapid access to data from both *in situ* and satellite-based observations.

Until then, statistical models will be used to predict where and when HABs are likely to occur based on historical records of the location, frequency and magnitude of HABs or on correlations of HABs with environmental variables or indices. This not only places a high priority on research (predictive models, real-time sensing technologies), it places a high priority on detection, i.e., initially, GOOS must focus on the development of the capacity to detect HA species and toxins routinely and rapidly in the context of changes in the distribution of key environmental factors. To these ends, priority should be placed on the establishment of a global network of sentinel (early warning) and reference (to develop climatologies of HABs and associated environmental conditions) stations for long-term time series observations and the development of an integrated data communications and management system for rapid access to and dissemination of data on HA species abundance, toxin concentrations, and key environmental variables (temperature, salinity, surface waves and currents, concentrations of inorganic and organic forms of N, P and Si). "Hot spots" where blooms are most likely to occur and have the greatest impact should be targeted.

5.7 REFERENCES

Barber and Chavez, 1986; Wilkinson *et al.*, 1999; Arcos *et al.*, 2001; Koblinsky and Smith, 2001; Lipp *et al.*, 2001; Beaugrand *et al.*, 2002; Francis *et al.*, 1998

Barry, J.P., C.H. Baxter, R.D. Sagarin, and S.E. Gilman. 1995. Climate-related, long-term faunal changes in a California rocky intertidal community. Science, 267: 672-675.

Botsford, L.W., J.C. Castilla, and C.H. Peterson. 1997. The management of fisheries and marine ecosystems. Science, 277: 509-515.

Carlton, J.T. 1996. Marine bioinvasions: the alteration of marine ecosystems by nonindigenous species. Oceanography, 9: 36-45.

Cheves, 2003 Cheves, M., 2003. Monitoring the heartbeat of mother earth. EOM, 12 (6): 6-10.

Cicin-Sain, B. and R.W. Knecth. 1998. Integrated Coastal and Ocean Management. Island Press, Washington, D.C., 517 pp.

Clark, J.S., S.R. Carpenter, M. Barber and others. 2001. Ecological forecasting: an emerging imperative. *Science*, 293: 657-660.

Costanza, R., W.M. Kemp and W.R. Boynton. 1993. Predictability, scale, and biodiversity in coastal and estuarine ecosystems: implications for management. *Ambio*, 22: 88-96.

Cowell, R. 1996. Global climate and infectious disease: the cholera paradigm. Science, 274: 2025-2031.

Dufour, A. P., 1984, Health Effects Criteria for Recreational Waters, Technical Report; U.S.E.P.A. Washington, DC. EPA 600/1-84-004

Epstein, P.R. (ed) 1999. Extreme Weather Events: The Health and Economic Consequences of the 1997/98 El Niño and La Niña. The Center for Health and the Global Environment, Harvard Medical School, Boston, Massachusetts, 41 pp.

Executive Order 13340, President George Bush, May 2004 Created a cabinet-level Task Force convening a Regional Collaboration of U.S. Federal, Great Lake States, local communities, Tribes and other interests in the Great Lakes Region and Canada.

Fisher *et al.*, 2003 Fisher, W.S., T.C. Malone, and J.D. Giattina. 2003. A pilot project to detect and forecast harmful algal blooms in the northern Gulf of Mexico. Environ. Mon. Assess., 81: 373-381.

Flather, R.A. 2000. Existing operational oceanography. Coastal Engineering, 41: 13-40.

Francy, D. S., and Darner, R.A., 1998, Factors affecting *Escherichia coli* concentrations at Lake Erie public bathing beaches. U.S.Geological Survey, Columbus Ohio, Water Resources Investigations Report 98-4241.

Francy, D.S., Gifford, A.M., and Darner, R.A. 2003. *Escherichia coli* at Ohio bathing beaches—distribution, sources, wastewater indicators, and predictive modeling. U.S.Geological Survey, Columbus Ohio, Water-Resources Investigations Report 02-4285. Full report available at *http://oh.water.usgs.gov/beaches/index.html*

Frick, W., Francy D.S., Rockwell, D.C., Schwab, D., Beletsky, D, and Lunetta, R, 2004, A Vision for a Beach Forecasting Tool, presented at MWWD 2004 – 3rd International Conference on Marine Waste Water Disposal and Marine Environment, and IEMES 2004 – 1st International Exhibition on Material Equipment and Services for Coastal WWTP Outfalls and Sealines, Catania (I)

GESAMP. 2001. Protecting the Oceans from Land-based Activities: Land-based Sources and Activities Affecting the Quality and Uses of the Marine, Coastal and Associated Freshwater Environments. GESAMP Reports and Studies No. 71, 162 pp.

Haines, A. and M. Parry. 1993. Climate and human health. J. Roy. Soc. Med., 86: 707-711.

Haines and Parry, 1993; Rogers and Packer, 1993; Colwell, 1996

Wade Jacobson, August 1997, Summary of U.S. Great Lakes Beach Closings 1981-1994, David Rockwell, Project Officer, EPA 905-R97-003

Hallegraeff, G.M., 1998. Transport of toxic dinoflagellates via ship's ballast water: bioeconomic risk assessment and efficacy of possible ballast water management strategies. Mar. Ecol. Prog. Ser., 168: 297-309.

Heinz Center. 2002. Human Links to Coastal Disasters. The H. John Heinz III Center for Science, Economics and the Environment, Washington, D.C. 139 pp.

Hofmann,E.E. and M. A. M. Friedrichs, 2002. Predictive Modelling for Marine Ecosystems. In: *The Sea*. Biological-Physical Interactions in the sea. A. R. Robinson, J. J. McCarthy, and B. J. Rothschild (eds), Volume 12, J.Wiley and Sons.

Hu *et al*., 2003 Hu, C., K.E. Hackett, M.K. Callahan, S. Andréfouët, J.L. Wheaton, J.W. Porter, and F.E. Muller-Karger. 2003. The 2002 ocean color anomaly in the Florida Bight: a cause of local coral reef decline? Geophys. Res. Letters, 30: (in press).

IOC (Intergovernmental Oceanographic Commission). 1998. The Global Ocean Observing System Prospectus 1998. IOC, Paris. 168pp. http://www.ioc.unesco.org/goos/Prospe98/Contents.html.

IOC (Intergovernmental Oceanographic Commission). 2003. The Integrated Strategic Design Plan for the Coastal Ocean Observations Module of the Global Ocean Observing System. GOOS Report No. 124, 190 pp. UNESCO (http://ioc.unesco.org/goos/docs/doclist.htm)

Jackson, J.B.C., M.X. Kirby, W.H. Berger, K.A. Bjorndal, L.W. Botsford, B.J. Bourque, R.H Bradbury, R.Cooke, J. Erlandson, J.A. Estes, T.P. Hughes, S. Kidwell, C.B. Lange, H.S. 2001. Historical overfishing and the recent collapse of coastal ecosystems. Science, 293: 629-643.

Kamykowski *et al*., 1998 Kamykowski, D., E.J. Milligan, and R.E. Reed. 1998. Biochemical relationship with the orientation of the autotrophic dinoflagellate *Gymnodinium breve* under nutrient replete conditions. Mar. Ecol. Prog. Ser., 167: 105-117.

Koblinsky, C. and N.R. Smith. 2001. Observing the Oceans in the 21st Century. GODAE Project Office, Bureau of Meteorology, Melbourne, Australia, 604 pp.

Kusek *et al*., 1999 Kusek, K.M., G. Vargo and K. Steidinger. 1999. *Gymnodinium breve* in the field, in the lab, and in the newspaper – a scientific and journalistic analysis of Florida red tides. In T.A. Villareal (Ed.), Contributions in Marine Science, 34: 1-229.

Lenihan *et al*. (2001) Lenihan, J.M. Pandolfi, C.H. Peterson, R.S. Steneck, M.J. Tegner and R.R. Warner. 2001. Historical overfishing and the recent collapse of coastal ecosystems. Science, 293: 629-643.

Levitus, S. J.L. Antonov, T.P. Boyer and C. Stephens. 2000. Warming of the world ocean. Science, 287: 2225-2229.

Malone, T.C., A. Knap, and M. Fogarty. 2005. Overview of science requirements. In Robinson, A.R. and K. Brink (eds.), *The Global Coastal Ocean: Multiscale Interdisciplinary Processes, The Sea*, Vol. 13, p. 757-784, Harvard University Press, 1033 pp.

McKay, B. and K. Mulvaney. 2001. A review of marine major ecological disturbances. Endangered Species Update, 18: 14-24.

Michaels, A., D. Malmquist, A. Knap and Ann Close. 1997. Climate science and insurance risk. Nature, 389:225-227.

Myers, R.A. and B. Worm. 2003. Rapid worldwide depletion of predatory fish communities. Nature, 423: 280-283.

Najjar, R.G., H.A. Walker, P.J. Anderson and others. 1999. The potential impacts of climate change on the Mid-Atlantic coastal region. Climate Res., 14: 219-233.

Nevers M.B., and Whitman R. L, 2004, Protecting visitor health in beach waters of Lake Michigan: Problems and opportunities in The State of Lake Michigan: Ecology, Health and Management, Edited by T. Edsall and M. Munawar Ecovision World Monograph Series, Aquatic Ecosystem Health and Management Society, 18

Nicholls, R.J. and C. Small. 2002. Improved estimates of coastal population and exposure to hazards. *Eos Transactions*, 83: 301-305.

NRC. 1999. From Monsoons to Microbes: Understanding the Ocean's Role in Human Health. National Academy Press, Washington, D.C. 132 pp.

Olyphant, G.A., Thomas J., Whitman R.L., and D. Harper 2003, Characterization and statistical modeling of bacterial (Escherichia coli) outflows from watersheds that discharge into southern Lake Michigan. Environ. Monitoring Assessment. 81, 289-300

Peierls, B.L., N.F. Caraco, M.L. Pace and J.J. Cole. 1991. Human influences on river nitrogen. Nature, 350: 386-387.

Rabinovici, S.J.M, Bernknopf R. L, Wein, A.M, Coursey, D. L, and Whitman R.L., 2004, Economic and Health Risk Trade-Offs of Swim Closures at a Lake Michigan Beach, Environ. Sci. Technol. 2004 38 (10), 2737-2745

Rogers, D.J. and M.J. Packer. 1993. Vector-borne diseases, models and global change. The Lancet, 342: 1282-1284.

Schofield *et al.*, 1999 Schofield, O., J. Grzymski, W.P. Bisset, G.J. Kirkpatrick, D.F. Millie, M. Moline, and C.S. Roesler. 1999. Optical monitoring and forecasting systems for harmful algal blooms: possibility or pipe dream? J. Phycol., 35: 1477-1496.

Sherman, K. and A.M. Duda. 1999. An ecosystem approach to global assessment and management of coastal waters. Mar. Ecol. Prog. Ser., 190: 271-287.

Small, C. and R.J. Nicholls. 2003. A global analysis of human settlement in coastal zones. J. Coastal Res. 19: 584-599

Steidinger *et al.*, 1997 Steidinger, K.A. 1975. Implications of dinoflagellate life cycles on initiation of *Gymnodinium breve* red tides. Environ. Lett., 9: 129-139.

Steidinger, K.A., G.A. Vargo, P.A. Tester, and C.R. Tomas. 1997. Bloom dynamics and physiology of Gymnodinium breve. In D.M. Anderson, A.E. Cembrella and G.M. Hallegraeff (Eds.), The Physiological Ecology of Harmful Algal Blooms, Elsevier, Amsterdam.

Stumpf *et al.*, 2003 Stumpf, R.P., V. Ransibrahmanakul, K.A. Steidinger and others. 1998. Observations of sea surface temperature and winds in association with Florida, USA red tides (*Gymnodinium breve*). In B. Reguera, J. Blanco, M.L. Fenandez, and T. Wyatt (Eds.), Harmful Algae. Xunta de Galacia and Intergovernmental Oceanographic Commission, UNESCO, Paris, France, p. 145-148.

Stumpf, R.P., M.E. Culver, P.A. Tester, M. Tomlinson, G.J. Kirkpatrick, B.A. Pederson, E. Truby, V. Ransibrahmanakul, and M. Soracco. 2003. Monitoring Karenia brevis blooms in the Gulf of Mexico using satellite ocean color imagery and other data. Harmful Algae, 25: 1-14.

Subramaniam and Carpenter, 1994 Subramaniam, A. and E.J. Carpenter. 1994. An empirically derived protocol for the detection of blooms of the marine cyanobacterium Trichodesmium using CZCS imagery. Int. J. Remote Sensing, 5: 1559-1569.

UNEP. 2001. Ecosystem-based management of fisheries: opportunities and challenges for coordination between marine Regional Fishery Bodies and Regional Seas Conventions. UNEP Regional Seas Reports and Studies, No. 175, 52 pp.

Vitousek, P.M., J.D. Aber, R.W. Howarth, G.E. Likens and others. 1997. Human alterations of the global nitrogen cycle: causes and consequences. Ecol. Appl., 7: 737-750.

Walsh, J.J. and K.A. Steidinger. 2001. Saharan dust and Florida red tides: the cyanophyte connection. J. Geophys. Res., 106: 11597-11612.

Whitman, R.L, 2005, Progress Report Grant Project Number GL98500001, City of Chicago Department of Environment, David Rockwell Project Officer

Whitman, R.L., Nevers, M.B., Gerovac, P.J., 1999. Interaction of ambient conditions and fecal coliform bacteria in southern Lake Michigan waters: monitoring program implications. Nat. Area J. 19, 166-171

6

A HACCP approach to manage water safety for large water basins

D. Deere and A. Davison

© 2008 IWA Publishing. *Effective Cross-Border Monitoring Systems for Waterborne Microbial Pathogens*. Edited by J. B. Rose and E. A. Dreelin. ISBN: 9781843391692. Published by IWA Publishing, London, UK.

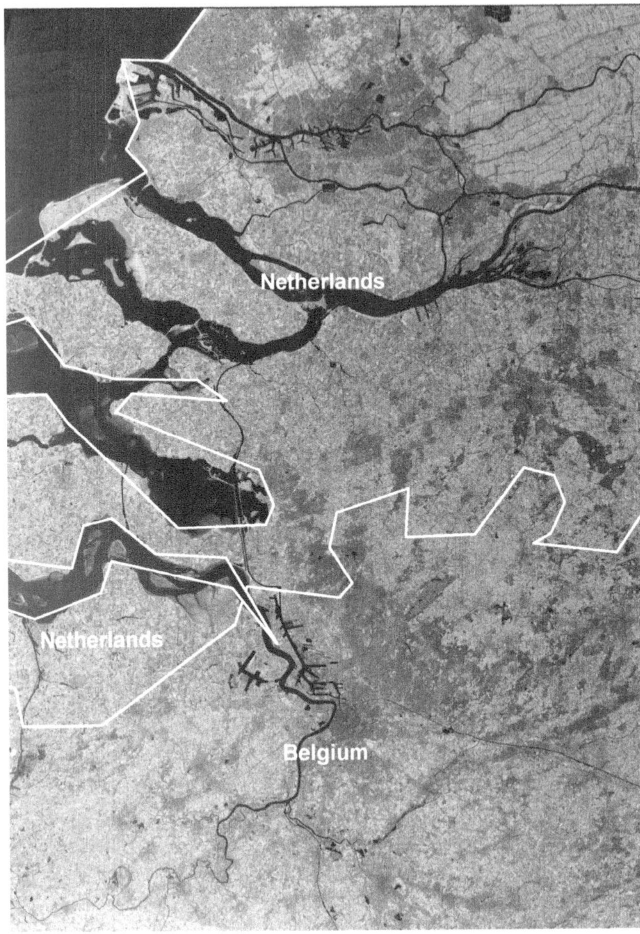

Rhine River. Image: MSU Center for Global Change and Earth Observations

6.1 INTRODUCTION

6.1.1 Application of HACCP to cross-border water resources management

The hazard analysis and critical control point (HACCP) system (WHO 2001) is widely used in the management of food and water quality and safety (Martel *et al.* 2006). At its simplest, HACCP consists of five preliminary steps and seven principles (Figure 6.1). In its contemporary form, however, HACCP is implemented as part of a risk management system that includes whatever supporting program are necessary to ensure reliable management of identified risks.

This chapter describes how the HACCP principles can be applied within the broader context of cross-border water resources management. The specific application described is the theoretical application of HACCP to the 5 different lakes and the Gulf of St. Lawrence Sea that are influenced by the surrounding states and provinces in the US and Canada.

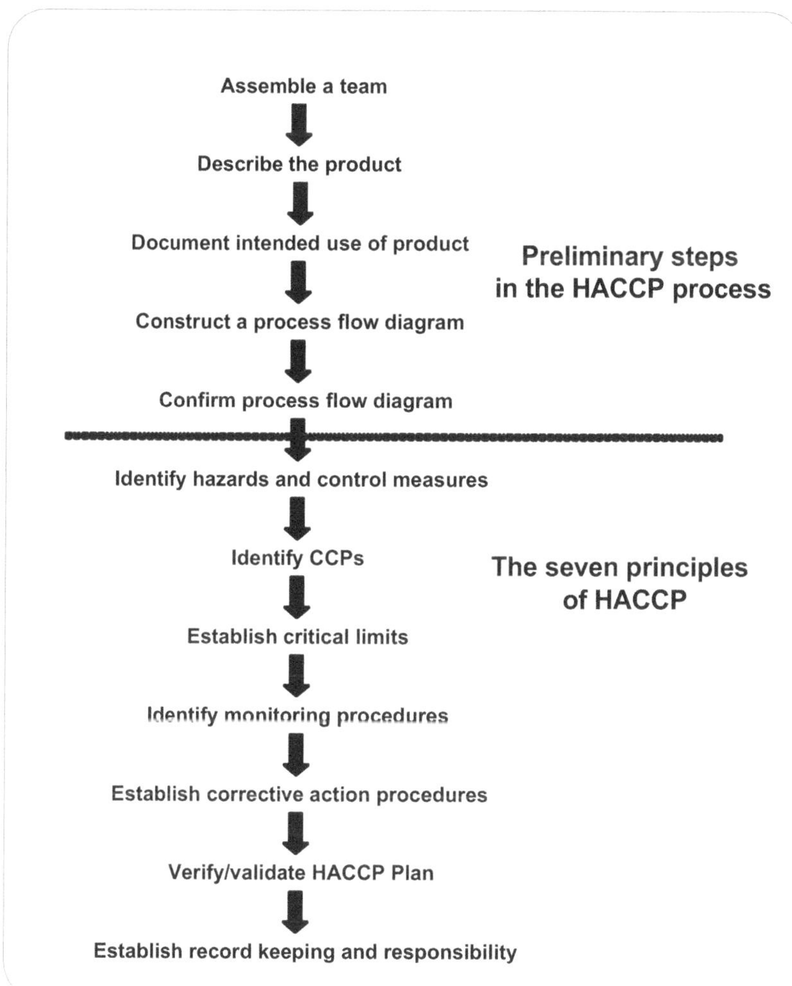

Figure 6.1. The steps and principles of HACCP.

There are numerous publications describing HACCP but the point of departure for HACCP guidance is that given in the Codex Alimentarius ('food code') (WHO 2001). Although promulgated under food safety management, the HACCP principles and concepts can be just as readily applied to the management of risks to water quality (Havelaar 1994, Deere and Davison 1998). One of the first practical applications of HACCP principles to watershed management was described by Barry *et al.* (1998) for the Californian Alameda Creek watershed that supplies drinking water to San Francisco. Since that time there have been numerous applications of HACCP principles to watershed protection.

In practice, it is common practice to tailor the way in which HACCP principles are applied to specific applications. For example, the New Zealand, Australian, International Water Association, AwwaRF and WHO drinking water charters, guidelines and regulations (reviewed by Deere and Davison 2005) are all recognizably analogous to HACCP, but do not necessarily use the acronym nor do they all describe their systems as being HACCP.

One of the reasons that the HACCP acronym is not used in water cycle management is that the 'jargon' used in HACCP, including the acronym itself, is often unfamiliar outside the health sector and the food industry. Communication of HACCP is sometimes aided by keeping all the principles but changing the terminology. Another reason for the departure is that organizations tend to develop their own systems under their own banners or brands rather than use an open acronym such as HACCP or some other system. Perhaps the most substantive reason is that there can be some difficulties in applying the more rigorous control approaches espoused by HACCP in situations where, in practice, control is achieved merely by influence, as in many watersheds.

In its contemporary application, HACCP has come to include any supporting programs that are required to manage the risk. For example, operational water safety management systems that are based around HACCP are typically supported by programs such as calibration, preventive maintenance, hygiene and sanitation, pest control, inputs and raw materials quality assurance, non-conforming water quality response, customer complaint response, training and awareness, awareness of legal requirements and "best practice" and a range of application specific "Operational Prerequisite Programs".

Most recently, the steps, principles and supporting programs of HACCP have been incorporated into a new international standard, "ISO 22000:2005 Food safety management systems – Requirements for any organization in the food chain." ISO 22000:2005 sets out a holistic framework, including Prerequisite Programs and Control Measures, for management of the whole supply chain – including upstream and downstream providers and suppliers.

Whereas there were sometimes difficulties in applying the 'critical control point concept' to watersheds, the Operational Prerequisite Programs defined within ISO 22000:2005 are readily interpreted for watersheds. Within the watershed context, the Operational Prerequisite Programs would include programs such as land use planning, pollutant load modelling, water quality testing, best management practice programs, education programs and sanitary inspection.

A remaining issue is that ISO 22000:2005 is defined in terms of the food chain. Whilst food chain includes drinking water, fisheries and stock and irrigation water, there are broader environmental values to be managed in watersheds that are not directly part of the food chain. Conceptually this is not a problem as the principles of systematic risk management espoused within ISO 22000:2005 and HACCP can be applied to manage risks to ecological as well as food chain endpoints. However, it is likely that there will be some instinctive rejection of ISO 22000 by watershed managers who might feel uncomfortable with its food chain connotations regardless of how applicable the system might be.

When providing training to those unfamiliar with water cycle risk management systems, simple interpretations are often useful, with minimal jargon. For readers new to HACCP-speak, we now provide an overview of the simplest breakdown of risk assessment and management systems as applied to any given endpoint (including water quality) (Figure 6.2). Risk management involves asking three questions:

- What are the risks (*syn* threats, potential adverse outcomes) that need to be managed?
- How can I manage (*syn* mitigate, control) these risks?
- How can I be confident that the risks are under control?

Another way of saying this is to think of risks as being problems:

- What problems might arise?
- How will I know if there is a problem?
- What will I do if I detect the problem?

Note that risk management systems are cyclical, the objective being to continuously review (actually, the review is usually annual to triennial) and improve the implemented systems (as shown in Figure 6.2). The objective of the review is to facilitate the incorporation of new knowledge from research as well as improve the rigor of the risk management activity as greater experience is gained. The continuous review process is analogous to the Kaizen improvement process or the "Plan – Do – Check – Act" cycle commonly described in generic management systems.

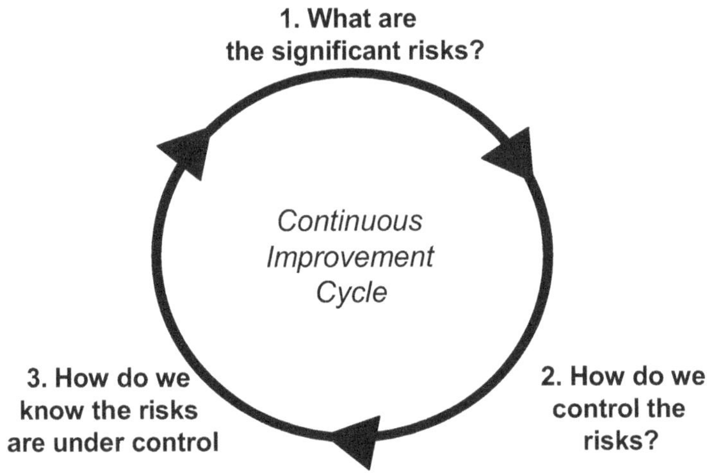

Figure 6.2. "HACCP Simple" (after Davison *et al* 2001).

6.1.2 Why create a HACCP-based risk management system?

The purpose of creating a risk management system is to document the major sources of risk to the endpoint of concern, identify and implement the major means of controlling those risks, undertake monitoring to provide early warning of the failure of those control processes and to have "ready to go" corrections that will be implemented if control processes fail. At the core of the risk management system is the HACCP plan which, if technically sound, and implemented properly in practice, should provide the structure within which to manage risks.

6.2 CREATING THE HACCP PLAN

6.2.1 The first step: Commit the resources to the task (Step 1)

The first step in developing a HACCP plan is to ensure a resource commitment from those that will create it. The resource commitment does not have to involve cash but must involve people-time. The following order of events would be recommended for the development of a large water basin HACCP plan, such for as for the Great Lakes:

1. The stakeholder agencies that oversee and regulate water resources management are identified and listed. Collectively, these agencies need to be able to identify risks and the means for the control of those risks;
2. A senior representative from each agency agrees in principle to the creation of a HACCP plan for the basin and identifies themselves as the 'HACCP Champion' for their organization. As the name implies, their role is to champion the development of the HACCP Plan from start, through to completion and thereafter to implement, maintain and update the plan over the long term;
3. The agency HACCP Champions identify a coordinator, whose job it is to coordinate and project manage the development of the HACCP plan;
4. Initially, the 'HACCP Coordinator' is fully dedicated to developing a HACCP implementation plan, essentially a conventional 'project management plan', but taking into consideration the fact that multiple agencies will be involved; and
5. The HACCP Champions collectively commit the required resources (at least human, and cash if required) to the development of the HACCP Plan. Individuals are named and portions of their time are committed according to the plan developed by the HACCP Coordinator.

As an example, for the Great Lakes, there are two lead groups of organizations: those from Canada, to be led by Environment Canada, those from the US, to be led by the US EPA. To avoid the perception that one or another is leading over the other, an independent third party could be appointed as the HACCP Coordinator. Alternatively, a two-person team could be set up and jointly function as a coordinating unit.

In practice, it is likely that the project management plan would consist of two parts. The first part would involve the scoping phase. The amount of work required to build, implement, maintain and update the HACCP plan cannot be predicted until the completion of the scoping phase.

One way to capture the identity of the assembled resources is to list names, roles and contact details, as in the example given in Table 6.1.

Table 6.1. Example of the HACCP Team committed to the Scoping Phase of the HACCP Plan for the Great Lakes Basin (content and footnotes are illustrative fictitious examples only)

Name	Host	Position	Primary role in HACCP team*	% Time	Duration	Email Details
			Jointly appointed HACCP Coordinator	100	July, 07 to June 30, 09	
			USEPA Champion	5	July, 07 to Sept 30, 07	
			Environment Canada Champion	5	July, 07 to Sept 30, 07	
			USEPA resource person	25	July, 07 to June 30, 09	
			Environment Canada resource person	25	July, 07 to June 30, 09	
Prof Joan Rose	Michigan State University	Homer Nowlin Chair in Water Research	Public health microbiology expertise	5	July, 07 to Sept 30, 07	
Kathy Martel	Cadmus Group Inc	Consulting Engineer	Water HACCP Trainer and Team Coach	25	July, 07 to June 30, 07	
Dr Melita Stevens	Melbourne Water Corporation	Manager, Drinking-Water Research	HACCP in water expertise	5	July, 07 to Sept 30, 07	
			Other agency representatives and technical experts as required...			

*Many on the team will play multiple roles and offer support in many areas. The purpose of this column is to denote the primary function fulfilled by the person named

6.2.2 The Scoping Phase: Explicit identification of water uses, users and their requirements (Steps 2 and 3)

Once a team has been committed and funding is in place for any travel and technical or professional support, the scoping phase can begin. HACCP is implemented as part of a risk management system. Therefore, the first part of the HACCP planning involves answering the following questions:

- To what and to whom are risks going to be assessed?
- What are the water quality requirements to meet the intended and incidental uses?

Both positive and negative affirmations are important. An inventory of water uses and users is required, along with their status as being either in or out of the plan. Against each use, the user expectations of water quality must be defined (see example in Table 6.2). In practice, it is common to start with just a few key water uses and users for the first iteration of the plan. Additional uses and users can be considered in future, expanded versions of the plan. This helps to focus the team initially and support more rapid completion of the plan. For example, Barry *et al* (1998) started with just one pathogenic hazard in their HACCP plan but over time added nutrients and sediments. To avoid perceptions of bias or of unethical treatment of certain user groups, it is important to state clearly the basis for inclusion or exclusion of groups. For example, it is common to initially focus on priority areas such as the involuntary risks to public health and then start to bring in voluntary risks as well as environmental and amenity risks in subsequent iterations.

Table 6.2. Example of the use, user and quality scoping sheet for the Great Lakes Basin HACCP Plan (content and footnotes are illustrative fictitious examples only)

User	Use	In/ Out*	Quality requirements**
Detroit Water Works	Drinking Water	In	The prevailing USEPA maximum contaminant levels (MCL) for organic pollutants (reference…) < 2,000 *E. coli* per 100 ml 95%ile
Canadian Salmon Farmers Inc	Aquaculture	In	The prevailing USFDA MCLs for organic pollutants and heavy metals (reference…) The prevailing Health Canada MCLs for antibiotics (reference…) < 150 *E. coli* per 100 ml 80%ile
St Lawrence Motor Sports Association	Jet skiing	Out	
And others			

*The reason some users are out is purely to enable the first iteration of the HACCP plan to be completed in a timely manner. The first iteration of the plan will focus on the public health risks of the food and water supplied from the water resource to the general community where involuntary risk exposure may arise. Future iterations may consider voluntary exposures and will consider environmental and amenity risks as well as public health.

**In the table it is common to cite any lengthy water quality requirements rather than list all the specific requirements. These citations are then given a formal reference, either an internet site address or a document reference, or both. The use of citations helps to avoid lengthy repetition in the plan as well as helping to keep the plan up to date since even if the cited document is updated, the plan can remain current.

6.2.3 Developing a conceptual model of the system from a water safety perspective (Steps 4 and 5)

Once the uses and users that are to be considered in the HACCP plan have been explicitly identified, their place in the water quality system needs to be represented in a form that enables possible sources of risks to be identified along with possible controls. This step is one that needs to be undertaken diligently as the omission of major process steps would be likely to lead to water quality risks, or points where control can be applied, being missed.

It is important to think conceptually and use a flow diagram to represent the flow of water and contaminants. Information, such as that generated from modeling wind-driven movement, can be used to help work out where water flows to and from and, therefore, where pollutants might travel. At this step, however, a detailed analysis is not required, this step is mostly about identifying the process steps and the likely hazardous events that may occur there.

In practice, it is common to build multiple conceptual flow diagrams for a complex system. One set of flow diagrams might describe the hydrologic flow of water and the co-transport of pollutants from sources, through possible control points, to the points where users come into contact with the water. Another set of flow diagrams might describe the generic water quality processes that take place across the system. An example of a very simple conceptual flow diagram is given in Figure 6.3, which illustrates the conceptual water quality processes by which fecal pathogens from farms can reach a water supply offtake. A conceptual high-level overview of the Great Lakes system is given in Figure 6.4.

Once flow diagrams have been developed, it is important to verify their accuracy. To do this, members of the HACCP team may need to travel by ship and by air and gather comment from those with good system knowledge, to inspect the watersheds contributing to the water resources as well as the water resources themselves. Once factually verified, the flow diagrams should be endorsed by an accountable person (usually through sign off from the HACCP Coordinator).

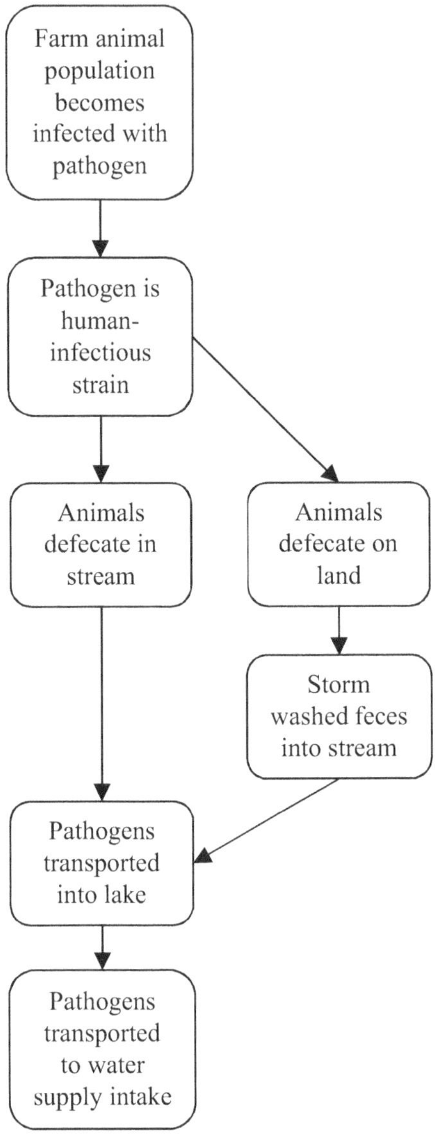

Figure 6.3. Simple conceptual process flow diagram illustrating the processes by which fecal pathogens from a farm might reach a water supply intake.

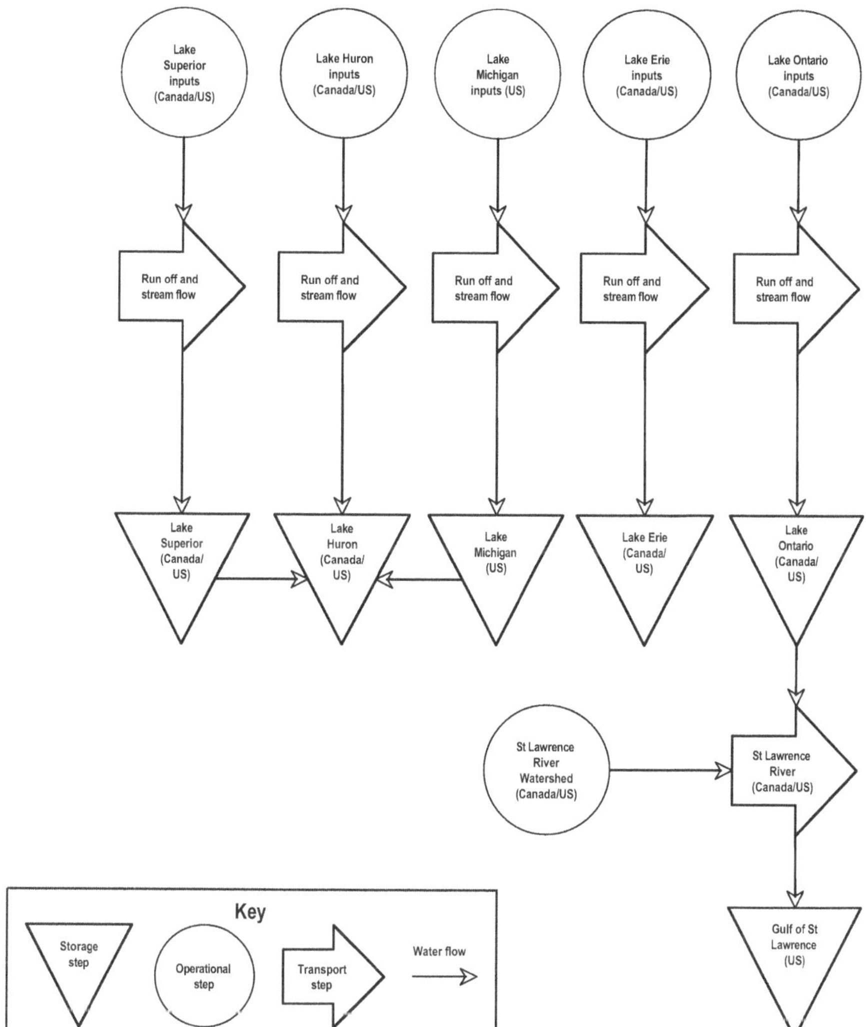

Figure 6.4. Conceptual high-level process flow diagram of the Great Lakes system.

6.3 PRINCIPLES

6.3.1 The Risk Assessment Phase "What are the risks?" (Principle 1)

With a conceptual understanding of the system in place, and a clear enunciation of the users and uses of the water resource, it is possible to start the process of systematically identifying the risks arising (Deere *et al* 2001). A risk needs to be a risk to something. In this case, the risk being assessed is:

- The risk of the water quality being unacceptable to the end user for the intended use.

The risk needs to arise from something. In this case, the origins of risk are assessed as:

- The mechanistic cause (the event, chain of events or scenarios) and the specific contaminating hazard (chemical, microbial or physical) that impairs water quality.

To make this component systematic, each step of the process flow diagram/s is/are analyzed to determine what could go wrong at that step to cause the water to become unfit for its intended use. Once hazards and causes have been identified, it is necessary to prioritize the most important through the risk assessment process. In practice, risks are mitigated by controls. Therefore, the control measures in place at each step are also identified to help establish what options are in place to reduce the risks to acceptable levels. If there are inadequate controls in place, additional controls need to be implemented, or the end uses need to be changed (Table 6.3).

In undertaking the risk assessment, it is advisable to be as quantitative as time and budg*et* allows. If there is very limited time and budget, it is better to use the expert judgment of the HACCP team and get started rather than delay endlessly awaiting more information ('Pareto principle' vs 'analysis paralysis'). The HACCP plan can be updated in future years as more and better data, information and knowledge is gathered from sanitary inspection, inventory, monitoring, modeling and scientific inference.

Table 6.3. Risk and hazard analysis example. In these examples, it is the risks arising from major storm events that are being assessed (content and footnotes are illustrative fictitious examples only).

Cause	Hazards	Maximum Risk	Controls	Residual Risk	Improvements
The event, chain of events or scenario that could lead to unacceptable water quality for the intended use from the perspective of the intended user.	The agent(s) that lead(s) to the water quality being unacceptable.	Level of risk if control measures fail.	Current control measures that are in place.	Level of risk if current control measures operate.	Additional control measures and/or research required or desired to better mitigate and/or understand the risk.
Major storm in Detroit causes stormwater pollution to overflow into Lake Huron and short circuit to the Port Huron Water Works intake.	Pathogens Organics	Significant for both hazards	Selective abstraction (for both hazards) Water treatment (for microbial hazards)	Insignificant for both hazards	Additional stormwater controls are desired since current controls are not 100% reliable. Additional research is required since the flow of hazards from the stormwater runoff to the intake is not properly characterized.
Sewage treatment plant bypass during major storm in Toronto causes ammoniacal intoxication of farmed salmonids in Lake Erie	Ammonia	Significant	Wet weather retention ponds	Significant, inadequate wet weather retention, deaths occur frequently	An increase in wet weather retention capacity is required. Research is required to identify the event magnitude at which dilution during the storm prevents intoxication during overflows.

6.3.2 The Risk Management Planning Phase "How do we manage the risks?" (Principles 2 to 5)

Once the risks and control measures have been identified, 'control loops' need to be described by which the operation of control measures can be monitored, failures picked up and responses put in place to protect the end users from impaired water quality. Control loops consist of critical limits, monitoring and corrections and are assigned to specific process steps that, under conventional HACCP terminology, are defined as critical control points (CCPs). CCPs are the points, steps or procedures at which control is applied. The purpose of listing CCPs is to help to focus attention on the specific conceptual process steps and/or points in space at which control can be applied.

Under the contemporary ISO 22000:2005 version of HACCP, the use of 'Operational Prerequisite Programs' (PRPs) is promoted to manage risks as well as CCPs. The operational PRPs are conceptually equivalent to CCPs but are those points or process steps where it is more difficult to develop precise control loops (critical limits/monitoring/correction). Naturally, watershed management is full of such controls and operational PRPs are likely to feature as much, if not more, than CCPs in contemporary watershed HACCP plans.

The precise difference between CCPs and operational PRPs is often very grey and it's important to just get on with the HACCP plan rather than being too 'semantic pedantic' about what the major control points in the process are called.

Some means of monitoring the effectiveness of the control measure(s) is required at each CCP or operational PRP. For each parameter monitored, a performance limit needs to be set, termed a critical limit in conventional HACCP terminology. In the event that monitoring reveals that a critical or operational limit has been exceeded at a CCP or operational PRP, a correction needs to be implemented to protect the end users from exposure to the impaired water quality. An example of control loops is given in Table 6.4. In practice, for both monitoring and corrections, the questions "why, where, when, who and how" should all be answered.

Table 6.4. Control loop examples. In these examples, the control loops are described for significant risks identified during the risk and hazard analysis (content and footnotes are illustrative fictitious examples only).

Cause of significant risk	**Hazards**	**Controls at CCPs or PRP**	**Monitoring** *The means of detecting the causal event and/or the failure of the control measures.*	**Critical limit** *The point at which the control measures fail and water quality might become unacceptable.*	**Corrections** *The means of protecting the end user from unacceptable water quality.*
Major storm in Detroit causes stormwater pollution to overflow into the Lake Huron and short circuit to the Port Huron Water Works intake.	Pathogens Organics	Selective abstraction at intake CCP (for both hazards) Water treatment at Water Works CCP (for microbial hazards)	Turbidity at intake point monitored on line alarmed to duty officer. Turbidity of water at filter outlet monitored on line alarmed to duty officer.	> 20 NTU for > 30 min > 0.3 NTU for > 15 min	Stop drawing from lake water and switch to stored water reservoir. Shut down water treatment works and switch to bores.
Sewage treatment plant bypass during major storm in Toronto causes ammoniacal intoxication of farmed salmonids in Lake Erie	Ammonia	Wet weather retention CCP	Level gauge monitored on line alarmed to duty officer.	> full depth for > 1 hr	Assess quantity of sewage overspilt. Compare to lookup table derived from historical data to see if ammonia level is likely to be toxic. If so, notify fish farm to enable fish to be harvested.

6.3.3 The validation and verification phase "How do we know the risk is managed?" (Principles 6 and 7)

The implementation of the HACCP plan begins once the planning phase is complete. There are several stages to implementation and included among them are some very testing, objective evaluation actions.

6.3.3.1 Validation

The HACCP plan relies heavily on scientific and technical understanding. If this understanding is incorrect, the HACCP plan is not valid. Therefore, a process of establishing validation of the HACCP plan needs to be undertaken largely before, but often including additional studies during, the HACCP plan implementation phase (Table 6.5). Validation relies on expert judgment, monitoring results and theoretical understanding.

Essentially, all of the important technical and practical aspects of the HACCP plan need to be explicitly described. The basis for the assumptions and choices made needs to be documented. In some cases, research will be required and an interim validation will need to be recorded.

Table 6.5. Validation schedule example. In these examples, the control loops are validated with reference to local observations and best practice guidance manuals (content and footnotes are illustrative fictitious examples only).

Item validated	Basis
Critical limit for turbidity for Port Huron Water Works intake point being > 20 NTU for > 30'.	Long term monitoring shows that pathogen and organic concentrations are low when turbidity is below 40 NTU. Treatment plant is able to run at full capacity up to 25 NTU.
Critical limit for turbidity at Port Huron Water Works filter outlet being > 0.3 NTU for > 15'.	Partnerships for Safe Water treatment guidance advises that turbidity should be < 0.3 NTU if water is sourced from compromised water source.
Toronto Sewage Treatment Plant retention pond overflow of > 1 hr.	Complaints of fish kills have only been recorded after spills have continued for more than one hour.

6.3.3.2 Verification

Once implemented, there needs to be two types of verification undertaken. The first type concerns verification that water quality requirements are met and includes water quality monitoring. The second type involves verification that the HACCP plan is being complied with in practice and includes auditing.

Auditing is not a popular subject area and can be redefined as 'assessment' or 'review' to avoid the negative connotations that the word 'auditing' brings with it from the financial auditing world. The focus of the auditing would be on the activities that are undertaken to control water quality (Table 6.6). For example, an audit might be used to establish that the monitoring of critical limits actually takes place, in the field, and that responses to nonconformities are appropriate.

Table 6.6. Verification schedule example. In these examples, the verification includes both audit-type and monitoring-type verification (content and footnotes are illustrative fictitious examples only).

Verification	Schedule
The water quality at the Port Huron Water Works intake point.	Monitoring of microbial pathogens and indicators at monthly intervals and daily following storm events until turbidity drops below 10 NTU. Monitoring volatile and semi-volatile organics quarterly and weekly following storm events until turbidity drops below 10 NTU.
Compliance with the calibration, maintenance, monitoring and response requirements for the level gauge used to detect overflows from Toronto Sewage Treatment Works retention pond.	Auditing of the calibration, maintenance and monitoring and response records at Toronto Sewage Treatment Works.

6.4 RECORD KEEPING

Records need to be kept, either electronic and/or documented, to enable trends to be analyzed and to help support auditing-type verification. This also helps to build up an evidence trail to support any legal defenses and demonstration of due diligence (Davison 2004). Ideally, records will be easy to interrogate and use in trend analysis and reporting.

6.5 SUPPORTING PROGRAMS

There are many activities and programs that are required to make a HACCP plan work. This is why HACCP in food safety management is just one component of a Food Safety Plan and HACCP in drinking water and water recycling management is just one component of a Water Safety Plan (Davison *et al* 2004). In this example, the HACCP plan needs to be supported within the broader integrated water resources management plan. A number of important supporting programs that need to exist in support of the HACCP plan include:

- training and educating those that manage the control loops that control the risks;
- stakeholder and water user awareness and education campaigns;
- communication protocols between the various agencies to support ongoing development and maintenance of the HACCP plan;

- research and continual improvement programs to generate a more precise and better validated HACCP plan into the future;
- inter-agency emergency management plans to respond to major contaminated water incidents such as after storms, spills or blooms; and
- systematic operation and maintenance programs with standard operating procedures to cover the calibration, maintenance, operation and management of the control loops that control the risks.

In practice, the development of supporting programs will build on those existing already so there is no standard list that can be provided. Once complete, the HACCP team should have identified, formalized and implemented any supporting programs that are required to make the HACCP plan work in practice.

6.6 BENEFITS OF THE HACCP APPROACH

Once complete and implemented, the HACCP plan provides confidence that the major risks have been identified and that ongoing, operational controls are in place to manage the risks to water quality and to give early warning if water quality is likely to become impaired. Over time, the plan can be improved, both in terms of it theoretical basis, through research, as well as through expansion of its scope and the rigor of its implementation.

6.7 REFERENCES

Barry, S.J., E.R. Atwill, K.W. Tate, T.S. Koopman, J. Cullor, and T. Huff. (1998) Developing and Implementing a HACCP-based Program to Control Cryptosporidium and Other Waterborne Pathogens in the Almeda Creek Watershed: Case Study. In Proc. of the AWWA Annual Conference. Denver, Colo.: AWWA.

Davison, A., Howard, G., Stevens, M., Callan, P., Kirby, R., Deere, D. and Bartram, J. (2004) Water Safety Plans. Protection of the Human Environment. Water, Sanitation and Health. WHO/SDE/WSH/02.09.

Davison, A.D., Schneider, P., Langdon, A., and Toop, P. (2001) Getting your priorities right: assessing and managing hazards. 19th Australian Water Association Federal Convention April 2001, Canberra.

Davison, A. (2004) Due Diligence Defense: The Benefits of a Documented System. HACCP in Water Utilities Conference, National Sanitation Foundation and World Health Organization, Ann Arbor, Michigan, 4-5 May.

Deere, D. and Davison, A. (1998) Safe water – are food guidelines the answer? Water 25:21-24

Deere, D., Stevens, M., Davison, A., Helm, G. and Dufour, A (2001) Management Strategies. In:

Water Quality: Guidelines, Standards and Health. Assessment of risk and risk management for water-related infectious disease. Fewtrell L. and Bartram J. (eds). IWA Publishing, London. ISBN 1 900222 28 0. Ch 12. p 257-288

Deere, D. and Davison, A. (2005). The Ps and Qs of Risk Assessment. Water 32(March): 38-43.

Havelaar, A.H. 1994. Application of HACCP to Drinking Water Supply. Food Control, 5(3):145-152.

Martel, K., Kirmeyer, G., Hansen, A., Stevens, M., Mullenger, J. and Deere, D. (2006) Application of HACCP for Distribution System Protection. Awwa Research Foundation, American Water Works Association, United States Environmental Protection Agency, International Water Association. AwwaRF Denver, CO, USA.

WHO (2001) Codex Alimentarius Food Hygiene Basic Texts. Food and Agricultural Organization of the United Nations, World Health Organization, Rome, 2001.

7

Transbounday Monitoring of Large Basin Systems for Waterborne Microbial Pathogens

J. B. Rose

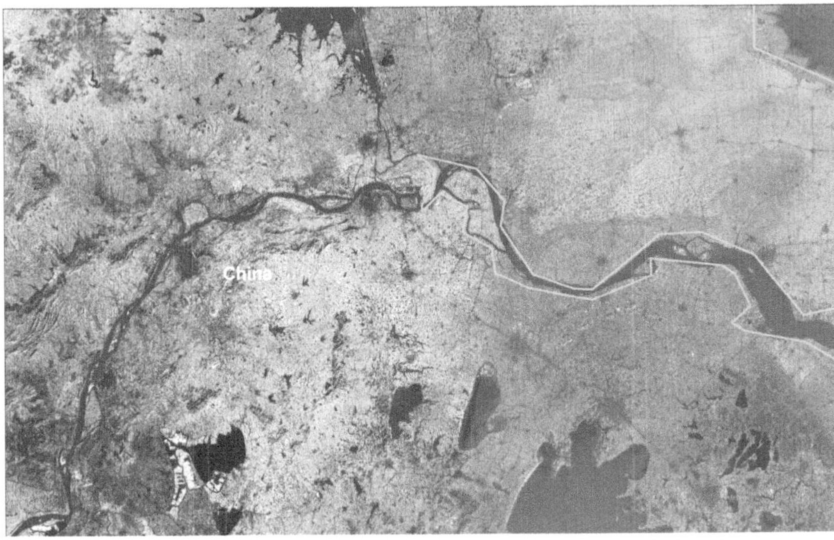

Yangtze River. Image: MSU Center for Global Change and Earth Observations

"Even diseases have lost their prestige, there aren't so many of them left. Think it over....no more syphilis, no more clap, no more typhoid....antibiotics have taken half the tragedy out of medicine"

-Louis Ferdinand Celine physician and poet 1894-1961

"The most fundamental refutation of the false doctrine of drinking-water epidemics is to be found in the works of Wolter. He and Emmerich have finally liquidated the drinking-water hypothesis"

-Georg Sticker 1912 (Howard-Jones, 1973).

"Of all the social and cultural crises we humans face, the water crisis is one that lies at the heart of our survival and that of our planet Earth."

-K. Matsuura, Director General, UNESCO (2003)

7.1 INTRODUCTION

7.1.1 Microbial Threats, Society and Science: Why We Need to Monitor our Water

History is filled with examples of the convergence of science and societal needs. None is more compelling and fascinating than the discovery and articulation of the causes of waterborne disease epidemics. The year was 1884, and Robert Koch, leader of the German Cholera Commission and arguably the most important scientist of his time, began addressing the issue of the contagious bacterium *Vibrio*. He was able to isolate the bacterium and contended that it was the cause of much of the pestilence and massive disease outbreaks throughout the world. In that same year, the Typhoid bacillus was isolated by Georg Gaffky (Beck 2000). However, John Snow and William Budd suggested much earlier (1849 and 1856; respectively) that cholera and typhoid fever were spread by feces and by water (Beck 2000). The evolution of the knowledge on waterborne pathogens followed several principles:

- There was a health-based target (in those early days it was focused on cholera and typhoid)
- A scientific method was used to identify "contagious agents" using theories developed by Robert Koch (1876) that guided the study of cause-and-effect under a science-based framework.
- Advancements in tools and techniques provided the means to undertake these investigations. These advancements included invention of the culture technique using salts and yeast in 1872, plating techniques in 1881 using gelatin, the use of sterilization (what is now known as the autoclave), and use of the microscope and Gram stains (and thus along came *Escherchia coli* [1884-1885]).

Despite early scientific studies that demonstrated waterborne transmission of cholera and typhoid along with the isolation of the bacteria from water and feces, a debate continued to rage between scientists regarding the "theory" that disease could be transmitted via the contamination of water by a living organism. Howard-Jones (1973) wrote about this forgotten debate in the history of public health; at that time research focused on what was called sanitary science. Interestingly, a public trial spurred the acknowledgement that water could spread disease. The trial dealt with the contamination of drinking water and an outbreak of typhoid in Gelsenkrichen, in the Ruhr valley in Germany in 1904. Robert Koch served as a scientific witness for the prosecution. The outbreak had occurred in 1901 and caused 3,000 individuals to become ill, 8% of which died and thus began a criminal prosecution of those responsible for the public works (Howard-Jones 1973). The trial revolved around several key points:

- A German Law (1879) made it a criminal offense to sell foodstuffs that could be injurious to health and offer a threat to life
- Water could be considered a "foodstuff" that is acquired and produced for public consumption.
- That as the level of the water supply became inadequate during certain times, particularly for industrial needs. The water company opened an emergency pipe to increase water levels and as a result untreated river water was distributed to the community.

Amazingly, Rudofl Emmerich, who at the time was Chair of the University Of Munich School Of Hygiene, served as a witness for the defense. He suggested that waterborne disease was not proven. However, the final outcome of the case was an affirmation by the court that the water company had "adulterated the filtered drinking water supply with untreated water" (Howard-Jones 1973).

Meanwhile in the US and Canada, waterborne typhoid outbreaks as well as dysentery were ravaging the populations living in the Great Lakes Basin (Wolman and Gorman 1931). And in 1914, as described by Dreelin in Chapter 3, the largest international transboundary water quality monitoring study was planned and implemented. If one is to examine this study closely, it would serve as a partial model for all of the hundreds of international water basins described in Chapter 2. The key principal components were:

1. Specifically addressing the status of water pollution that was injurious to public health.
2. Using maps to describe the full extent of the water basin and monitoring results.
3. Choosing sampling sites to represent a large spatial scale and key areas of concern.
4. Using the most advanced microbiological method of the time to study water quality, with the necessary laboratories equipped with the required supplies and trained personnel.
5. Studying the sources of pollution and exposures through the drinking water supply chain.

6. Developing the beginnings of a water quality agreement through which the various parties would address water issues of common interest at the local scale.

Waterborne disease continues to threaten both the developed and developing regions of the world. In developed countries, outbreaks are the "plane crashes" for the water industry, and underreporting of endemic waterborne disease has lead to complacency by the water industry. As a result, the populace as well as the government takes "safe" water for granted. Sattar and Tetro (Chapter 4) have fully described the current threats to our waters and changes in our communities. Increases in population size, and particularly increases in sensitive subsets of people such as those with immunocompromised states, highlights the importance of maintaining a safe water supply because the community water supply has the potential to serve as a predominant route of exposure. This could result in a significant impact to human health on a large spatial scale. The increasing reports globally of polio, cholera, antibiotic resistant typhoid, *Cryptosporidium* and *E.coli* 0157:H7 demonstrate not only the emergence of new pathogens but that fecal wastes from the growing numbers of humans and animals are impacting waters used for drinking and recreating (Rose and Farrah 2007).

In addition to changes on the land and in human population size, climate change impacts the distribution, transport and survival of waterborne pathogens and thus waterborne disease risk. Rainfall, runoff, and temperature are environmental factors that influence water quality and the transport of waterborne pathogens of fecal origin from their sources to exposure sites. Beyond the traditional waterborne pathogens associated with enteric bacteria, viruses, parasites, the hazardous algal species are now emerging as a considerable threat to human health (see Malone and Rockwell, Chapter 5). While we have relied on bacterial indicators to address water quality, it has been known and discussed for the last 50 years that this monitoring system alone can not be used to address the risk of waterborne disease. Important lessons are to be learned from the early work of Koch and other studies on water safety. A public health target must be the goal, with a better understanding of the key pathogens of concern, their occurrence in water and people, along with cause and effects. With the convergence of discoveries and advances in microbiology, medicine and engineering, an improved monitoring program is needed. We can no longer rely on the technology of the 19th century. Improved information on water quality at the appropriate scale will help communities prioritize their pollution problems and help move the political will toward broad scale protection of water basins and toward an investment in the integration of sanitation and water supply.

The World Health Organization (WHO) recognizes that a multi-barrier approach and watershed protection is necessary to achieve water safety. Thus the development of Water Safety Plans under the drinking water guidelines use the concepts developed for HACCP as a management strategy for water systems (See Deere and Davison, Chapter 6). Key components of this approach are the identification of hazards, critical risk points in the systems and those that are controllable, and ways to verify that the controls work.

This chapter will describe approaches that can be used to develop monitoring programs using new tools and technologies for addressing microbial water quality that might serve a number of purposes. These purposes include monitoring to meet standards via compliance monitoring. However, monitoring should also provide information for a number of other applications including development of microbial risk assessments tied to Water Safety Plans. Monitoring thus may provide information regarding:

- Characterization of sources;
- Assessment of threats (e.g., emerging contaminants; flooding);
- Evaluation of impact at exposure sites; and
- New methods for treatment and solutions.

These supplemental/knowledge based monitoring programs would be basin specific and address societal needs and emerging concerns as well as laboratory capacity and human resources. The information collected as a part of these monitoring programs should also feed into a global database on water quality, concerns, and solutions, and a global network for knowledge sharing.

7.2 TRADITIONAL MONITORING APPROACHES FOR MICROORGANISMS

7.2.1 The Current Water Standards and Guidelines for Microorganisms

The safety of drinking water associated with microbial contaminants historically has been viewed from the perspective of eliminating "outbreaks" and observable illness in populations using surveillance and epidemiological studies. Early surveillance was primarily focused on bacterial waterborne disease associated with cholera and typhoid and fecal contamination of water. The focus on bacteria led to the development of the use of fecal indicator bacteria and the use of total coliform bacteria. This approach of using fecal indicator bacteria had two major results:

1. Because there was an assumption that all surface waters used for drinking water supplies were "unsafe" without treatment, monitoring began to focus on the final product after treatment (filtration and disinfection). The goal became zero detection of total coliform bacteria in finished water because total coliforms were seen as more wide-spread and a supposedly hardier group than *E.coli*. Thus, a treatment indicator and monitoring program was implemented to establish the goal of ZERO total coliforms in 100 ml of treated water. Australia, European Union and the WHO no longer require total coliform monitoring and use only *E.coli* where as Canada, uses the total coliform bacteria for operations evaluation as mentioned above.

2. *E.coli* however, has become the global standard for fecal indicator bacteria. It has been suggested that *E. coli* has a greater association to fecal contamination and to health risk in humans than total coliforms. For example, in the US boil water orders must be issued if *E.coli* is present but not if only total coliforms are found. *E. coli* is also used to assess quality and safety of groundwater resources.

There have been no studies that actually demonstrate a dose-response relationship between indicators and public health risk for drinking water for either surface waters with treatment or groundwaters. Thus, monitoring of drinking waters has focused on treatment and the assurances that treatment works or in some cases that groundwater is protected. These monitoring programs produce information on compliance, but have not really worked to produce an assessment of risk. Hundreds of boil orders are issued in the developed regions of the world. These boil water orders and the numbers of cholera and typhoid in the developing world are a testament that drinking water safety is not being achieved. NOTE: The current disease burden of typhoid is estimated at 17 million per year and 600,000 deaths while cholera is reportable and the numbers range from 100,000 to 250,000 cases per year with about 2000-3000 deaths. (WHOhttp://www.who.int/vaccine_research/diseases/typhoid/en/)
(WHO, http://www.who.int/wer/2005/wer8031.pdf).

Most countries state a goal of zero pathogens detected in treated water in their drinking water standards. However, drinking water monitoring programs for specific pathogens were slow to develop. Enteric viruses and the protozoa *Cryptosporidium* and *Giardia,* have been monitored in surface waters. Safety regarding these pathogens was addressed according to treatment technology rules, presumably to achieve a specific level of reduction of "any" of the pathogens in the source water via drinking water treatment. For surface water, this treatment requirement amounted to filtration and disinfection and for groundwater, disinfection. Monitoring for specific pathogens moved the use of microbial risk assessment methods (see section below) to address source water quality and treatment needs in the US, and the European Union, Australia and South Africa (Medema et al. 2006; Rose et al. 1991 Regli et al. 1991). Some of the rules in the US developed by the EPA for microbial contaminants are listed in Table 7.1.

Table 7.1. Examples of Rules within USEPA for Drinking Water Safety

RULE	MICROBES TO MONITOR	GOALS	TREATMENT	ISSUES
Surface Water Treatment Rule SWTR	*Giardia* & Viruses in surface waters *Legionella* not monitored for.	Reduce levels from surface water to a 10^{-4}, Annual risk of infection.	99.99% (4-log10) reduction of viruses 99.9% (3-log10) reduction of *Giardia* Developed disinfection and filtration credits.	*Cryptosporidium* was not addressed. Side stepped any real assessment or control for *Legionella* Did not address any other new pathogens.
Information Collection Rule ICR	*Cryptosporidium* *Giardia* & Viruses In suface waters.	Monitor levels in source water, determine treatment reductions needs for microbes. Balance DBP formation with microbe reduction.	Used to establish national contamination levels and address reduction goals. Membranes and UV considerations. (due to chlorination resistance of *Cryptosporidium*)	18 months of monitoring Methods development Parasite viability
Enhanced Long-Term Surface Water Treatment Rule ELTSWTR	*Cryptosporidium* *Giardia* In surface waters	Meet the 10-4 Annual risk of infection. Determine treatment needs using the ICR. Require more monitoring for surface waters.	May need more than 99% reduction of *Cryptosporidium*. (crypto likely the driver) Membranes, UV and ozone addressed.	*Cryptosporidium* driving the risk due to it's resistance to chlorine. Thus if source waters are highly contaminated, greater levels of reduction are needed including membrane filtrations, ozone or uv disinfection.

Table 7.1. (continued) Examples of Rules within USEPA for Drinking Water Safety

RULE	MICROBES TO MONITOR	GOALS	TREATMENT	ISSUES
Ground water Rule	Viruses in groundwaters Special National surveys.	Meet the 10^{-4} Annual risk of infection	To guide Disinfection requirements May monitor for coliphage and other fecal indicators (eg enterococci).	No good models for transport for assessment of vulnerability. Pathogen monitoring may be needed in future.
Contaminant Candidate List (CCL)	Includes emerging bacteria parasites viruses, to be identified, theoretically for both ground and surface waters.	New list of microbes to be addressed for regulation.	Need to determine occurrence (need methods to do this) . Need to determine health risk & risk of waterborne disease.	No occurrence data bases are moving forward.

The safety of ambient recreational waters has also relied on fecal indicators, primarily fecal coliform (thermotolerant coliform group of bacteria), *E.coli,* or enterococci. Water quality standards associated with these bacteria, along with monitoring approaches, are in place to protect public health at many beaches around the world. Guidelines (recommended levels) and standards (legally enforceable criteria that have been set by a governing body) have been developed based on dose-response evidence from water quality testing and epidemiological data on health risk. The guidelines and standards establish concentrations of the indicators to guide appropriate use of the water for recreation (full body contact versus boating for example). Unlike drinking water, studies have demonstrated a relationship between the fecal indicator levels and human health risks associated with people recreating in waters.

Table 7.2 lists the WHO, European Union (EU) and United States Environmental Protection Agency (EPA) guidelines and standards.

Table 7.2. World Health Organization, US Environmental Protection Agency, and European Union guidelines for recreational waters

Indicator	Levels per 100ml	Interpretation	Agency
Enterococci	≤ 40 41-200 201-500 >500	Marine & Fresh Waters Very Good, risk <1% Good, risk 1-5% Fair, risk 5-10% Poor, risk >10%	WHO
Enterococci	<35 Geometric mean 104 Single sample max	Marine Waters Risks <1.9%	USEPA
Enterococci	<33 Geometric mean 61 Single sample max	Fresh Waters Risks <0.8%	USEPA
E.coli	<126 Geometric mean <235 Single sample max	Fresh Waters Risks <0.8%	USEPA
Enterococci	200 (95 percentile) 400 (95 percentile) 330 (90 percentile)	Fresh waters Excellent Good Sufficient	EU

Table 7.2. (continued) World Health Organization, US Environmental Protection Agency, and European Union guidelines for recreational waters

Indicator	Levels per 100ml	Interpretation	Agency
Enterococci	 100 (95 percentile) 200 (95 percentile) 185 (90 percentile)	Marine waters Excellent Good Sufficient	EU
E.coli	 250 (95 percentile) 500 (95 percentile) 500 (90 percentile)	Marine waters Excellent Good Sufficient	EU

Monitoring programs for drinking and recreational waters are the primary microbial monitoring programs throughout the world. The difference between monitoring programs for drinking and recreational waters highlights the problems associated with setting up a large scale study:

- Programs use different indicators and time scales. Drinking water monitoring programs monitor the finished water using total coliform bacteria and *E.coli*, samples are generally collected daily to demonstrate the non-detects. Recreational water monitoring focuses mostly on enterococci or *E.coli* concentrations, samples are collected throughout the swimming season anywhere from daily to 3 to 5 times per month, with safe and risk levels denoted from dose-response functions between the indicator and disease. Both monitor exposure.
- Drinking water monitoring may include parasite monitoring in surface waters at the intake only. No pathogen monitoring has been used for recreational waters.
- For both drinking and recreational waters, the yardstick used to examine relative risks have been reports of actual waterborne disease associated with public health surveillance. As most surveillance systems are inadequate, this risk is not adequately assessed. Surface waters are presumed to be of moderate risk requiring treatment for drinking while one can bathe in those waters with various levels of risk dependent on the indicator levels. And finally, groundwaters are presumed to be low risk for drinking water; however, the data associated with outbreaks in both the developing and developed regions of the world demonstrate that this is not necessarily true (WHO 1999; Hrudey and Hrudey 2004) Figure 7.1

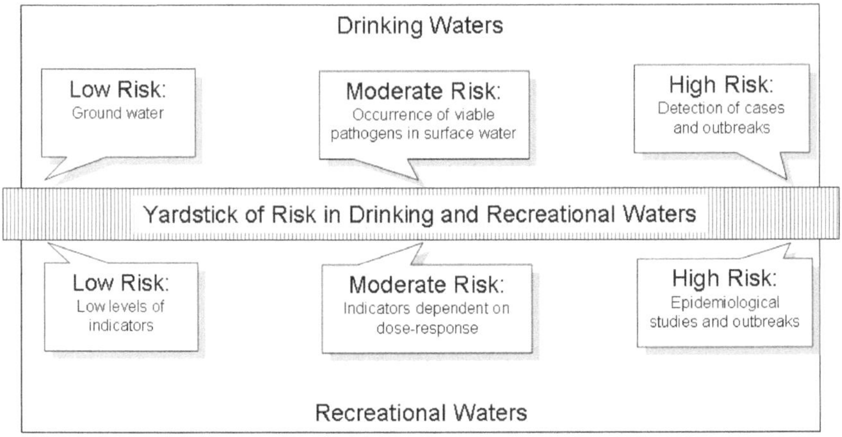

Figure 7.1. Assumptions used to address microbial risks for drinking water guiding treatment and recreational waters guiding use.

In addition, there are problems with an approach using only indicators because indicator monitoring:

- Does not address emerging microorganisms and those not associated with fecal contamination;
- Does not support a risk & management framework and a consensus toward a health-based goal;
- Does not adequately address treatment reduction and risks for parasites, viruses and other new waterborne risks (cyanobacteria);
- Does not address watershed protection (generally total reliance on drinking water treatment and addresses only exposure sites); and
- Does not address disasters and climate factors

7.2.2 The Indicators

There are a number of fecal indicators that can be used for assessing microbiological water quality. Using more than one indicator or alternative indicators provides better information, thus, it has been recommended that a tool box approach be used. In order to assess water quality in terms of microbial pollution, Fecal Coliforms, *Escherichia coli*, Enterococci are the main indicator microorganisms used and to some extent *Clostridium spp.* and bacteriophages (viruses that only infect bacteria) (Table 7.3).

Fecal Coliform bacteria belong to the Family *Enterbacteriaceae* (Table 7.3). Coliforms are aerobic and facultative anaerobic, non-spore forming, gram negative bacteria able to ferment lactose at 35ºC (Total coliform) and 44.5ºC

(Fecal coliform). Coliforms live in the digestive tract of warm-blooded animals (humans, pets, farm animals, and wildlife) and are excreted in the feces. In themselves, fecal coliforms generally do not pose a danger to people or animals, but they do indicate the possible presence of other disease-causing bacteria, such as those that cause typhoid, dysentery and cholera. *Escherichia coli (E. coli)* are a specific type of coliform bacteria that possess enzyme β-glucuronidase and are capable of cleaving substrate 4-methylum-belliferyl- β-D glucuronide (MUG). Their presence indicates specifically fecal contamination, and the possibility of enteric pathogens. *E. coli* is a worldwide universal indicator of fecal contamination of water. Some strains of *E. coli* are harmful and do cause illness. One type of *E .coli* (O157:H7) causes gastroenteritis (Levy et al. 1998) and has been the cause of both drinking and recreational disease. This serious pathogen has been found in water at beaches and is associated with illnesses in swimmers (Bruneau, A et al. 2004., Harrison and Kinra 2004, and Paunio et al. 1999).

Table 7.3. Classification of bacteria used as fecal indicators

Kingdom	Bacteria	Bacteria	Bacteria
Phylum (P) / Division (D)	Proteobacteria (P)	Firmicutes (D)	Firmicutes (D)
Class	Gamma	Bacilli	Clostridia
Order	Enterobacteriales	Lactobacillales	Clostridiales
Family	Enterobacteriaceae	Enterococcaceae	Clostridiaceae
Genus	*Escherichia*	*Enterococcus*	*Clostridiaceae*
Species	*Coli*	*avium* *durans* *faecalis* *faecium*	*Perfringens*
Shape	Rod	Cocci (sphereical)	Rod
Growth condition	Aerobic & Facultative anaerobes	Facultative anaerobes	Anaerobic
Gram +/-	-	+	+
Sporulate	No	No	Yes

Enterococci are part of what is known as the fecal streptococci group; *E. faecalis* and *E. faecium* are found in humans but are not restricted to humans. Enterococci can also cause nosocomial infections because they readily colonize skin and mucous membranes. These bacteria also tolerate temperatures from 10°C to 45°C, survive in acid and alkaline conditions and more resistant to environmental stress and chlorination than coliform bacteria and generally persist longer in the environment (Maier et al., 2000., Gleeson and Gray, 1997).

Clostridium perfringens are spore forming bacteria which live under anaerobic conditions. They are excreted in the feces, and can be an indication of old pollution as the spores are able to persist in the environment. Some have suggested that the persistence of spores makes *Clostridium* more suitable as an indicator of persistent pathogens of a fecal origin (Payment and Franco, 1993) such as viruses and parasites. Due to the ability of *E. coli* and *Enterococc*i to re-grow in the environment under warmer conditions whereas *C. perfringens* generally is not able to do so (Davies et al, 1995), *C. perfringens* is considered a pragmatic addition to other commonly used indicator species. Due to lower decay rates, *C. perfringens* can persist in the environment significantly longer than enteric pathogens (Cabelli, 1978) making them good indicators of fecal pollution (Ashbolt, 2001), especially in warmer tropical waters.

Viruses are microscopic nano-particles which infect cells of a living organism and cannot replicate on their own. Bacteriophage are viruses that infect bacteria, and coliphages are bacteriophages that are specific to the host *E. coli*. Because coliphages almost always come from fecal material, their presence in water bodies can indicate bacterial contamination as well as the survival and transport of human viral pathogens (Noble et al. 2003). Bacteriophages resemble enteric viruses in size, structure, morphology and behaviour in water (Pepper et al. 2006), therefore their presence in the environment can indicate a possibility of human viruses if sewage is the source of fecal contamination. Bacteriophages are also advantageous to use because they have not been shown to replicate in the environment.

7.3 ADVANCED MICROBIAL RISK FRAMEWORKS AND WATER QUALITY TOOLS

A revolution in genomics and molecular techniques in the field of microbiology has occurred over the past two decades. These new techniques have gained acceptance and standardization in the clinical field because rapid and accurate diagnosis was important for decisions regarding treatment of patients. Water quality monitoring has also begun to use these more accurate, specific tools and techniques to address sources of pollution, emerging microorganisms, health risks and priority management decisions. Molecular tools, particularly the polymerase chain reaction (PCR), will enhance our environmental monitoring programs. These tools and the information they provide should be used in quantitative microbial risk assessment frameworks, WHO Water Safety Plans, and HACCP.

7.3.1 The Microbial Risk Framework: Quantitative Microbial Risk Assessment

The WHO suggests that Water Safety Plans should establish a health goal focused on the prevention of waterborne disease through the identification of the hazards. The waterborne pathogen risk, their sources, transport/survival, and the potential to be exposed through water needs to estimated. Quantitative microbial risk assessment (QMRA) frameworks have been used to predict the likelihood of infection using monitoring data (Haas et al., 1999). Application of the QMRA framework 1) addresses specific microbial hazards, 2) uses of dose-response functions to examine exposure and outcomes, 3) applies quantitative estimates of the exposure through water, and 4) characterizes the probability of health outcomes including infection, morbidity and mortality. These health outcomes can be translated to DALY (Disability Adjusted Life Years, http://www.who.int/healthinfo/boddaly/en/) and can be used to address endemic and epidemic risks of waterborne disease.

Hazard Identification

Hazard identification determines which particular microorganisms are causally associated with waterborne disease and the types of health outcomes these organisms initiate. All fecally excreted microorganisms have some potential to be transmitted via water. Transmission is dependent on their concentration, survival and dose-response. Molecular technologies can help to identify specific microorganisms including viruses, parasites and bacteria; identify specific genes that are involved in the disease process as well as the severity of the disease; and aid in pathogen discovery. Molecular epidemiology has been used during outbreaks to examine the sources and pathways associated with the exposure. This will aid in prioritization in regard to establishment of health goals (eg. Reduction of typhoid in Africa and reduction of norovirus in North America) .

Dose-Response Assessment

Models for dose-response have been developed in laboratory studies with human volunteers or animals. The data are used to characterize the relationship between dose of a pathogen ingested/or inhaled and the incidence of a health adverse effect (i.e., the response). Two different dose-response models, the exponential and beta-Poisson models, are useful for QMRA. The dose-response relationship for the protozoa most commonly fits the exponential model whereas enteric bacteria and viruses, e.g. *Salmonella*, *Shigella*, *Vibrio*, *Campylobacter*, and *E. coli*, fit the beta-Poisson model (Haas et al. 1999). The models are:

$$P_i = 1 - e^{-rD} \qquad\qquad 1-\left[1+\frac{D}{N_{50}}\left(2^{1/\alpha}-1\right)\right]^{-\alpha} = \boldsymbol{P_i}$$

Where P_i is a probability of infection (ability of the organism to establish and reproduce in the intestine), D is an exposure dose expressed as numbers of the pathogen ingested, and α, N_{50}, k are constants for specific organism that define the dose-response in an Exponential or Beta Poisson model. Parameters for major waterborne pathogens are summarized in Table 7.4 and by Haas et al. 1999.

Table 7.4. Some Best Fit Dose-Response Parameters (from human experimental studies)

Organism	Exponential	Beta Poisson Parameters	
	k (r =1/k)	N_{50}	A
Poliovirus I (Minor)	109.87		
Rotavirus		6.17	0.2531
Hepatitis A Virus(#)	1.8229		
Adenovirus 4	2.397		
Echovirus 12	78.3		
Coxsackie (¶)	69.1		
Salmonella (**)		23,600	0.3126
Salmonella typhosa		$3.60 \bullet 10^6$	0.1086
Shigella (***)		1120	0.2100
Escherichia coli (%)		$8.60 \bullet 10^7$	0.1778
Campylobacter jejuni		896	0.145
Vibrio cholera		243	0.25
Endamoeba coli		341	0.1008
Cryptosporidium parvum	238		
Giardia lamblia	50.23		

(#) dose in grams of feces (of excreting infected individuals)
(*) multiple pathogenic strains pooled
(**) multiple (non-typhoid) pathogenic strains (*S. pullorum* excluded)
(***) *flexnerii* and *dysenteriae* pooled
(¶) B4 and A21 strains pooled
(%) Nonenterohemmohagic Strains (except O111)

Exposure Assessment

This process determines the concentrations and duration of pathogen exposure as well as the size and nature of the population that has been exposed. In water, this means considering the original load of the pathogens at the sources, attenuation by natural processes and reduction by engineered controls. This is the most difficult piece of the risk assessment and where much attention to monitoring programs in water should be paid.

Risk Characterization and Management

Based on information obtained from the exposure assessment for a specific pathogen and use of the specific dose-response equations, a quantitative risk can be estimated. Often, probabilities are developed based on distributions around the occurrence of the hazard, the variability in concentrations and the dose-response. Monte-Carlo simulations for analyzing uncertainty propagation are generally performed using computer software, (e.g. Matlab, R, and Crystal

Ball®), and help to determine how random variation, lack of information, or error affect the sensitivities or reliability of the analysis.

In order to integrate the monitoring programs using both traditional and innovative methods into a risk framework, the following is proposed:

- Hazard Identification would begin with traditional indicators and alternative indicators (particularly those that do not regrow such as *Clostridium* and coliphage) to provide a broad assessment of water quality and to identify hot spots regarding inputs into a watershed.
- The next monitoring tools that should be used for hazard identification would utilize microbial source tracking methods and finally actual pathogen monitoring using molecular and quantitative PCR techniques (see below). This would allow for strategic monitoring plans in time and place and using the more costly techniques only in areas with identified problems.
- Exposure assessment would use sanitary engineering data, land use, transport and survival studies associated with temperature, precipitation data, flows and run off to estimate exposure.

These new technologies will provide improved information in regard to public health threats and allow one to address the public health goals established within Water Safety Plans.

7.3.2 New Technologies

PCR

PCR is the most popular molecular technique to date. Any new pathogen can be detected using PCR once part of its genetic code is identified. PCR is an enzyme-driven method for amplifying short regions of DNA in vitro. PCR uses designed oligonucleotide primers based on known sequences of the target DNA a priori to hybridise specially to the target sequences. The target DNA is copied by a thermostable DNA polymerase enzyme, in the presence of nucleotides and primers. A specific microorganism can be detected by PCR via numerous cycles of heating and cooling in a machine known as a thermocycler to construct rounds of exponential increases in the target DNA (Mullis and Faloona 1987; Mullis 1990; Eisenstein 1990; White et al. 1992; Persing 1993; Yang and Rothman 2004). PCR has several advantages including:

- Live and dead microorganisms are detected, thus <u>potential</u> risks over time can be evaluated.
- Highly specific results are obtained generally in 24 to 48 hours.
- Microorganisms can be targeted which are difficult to cultivate or for which there is no cultivation procedure.

Quantitative Real-Time PCR (qPCR)

Probably the most recent advancement in molecular tools is qPCR, in which rapid quantification of the target genetic sequences can be made with amplification and detection combined in a single reaction vessel (Yang and Rothman 2004). qPCR techniques are rapidly developing with a variety of machines available and kits which allow for a standardized approach for detection of source tracking markers or specific pathogens. Overall advantages of qPCR compared with conventional PCR include not only quantitative capability but also speed, simplicity and reproducibility (Yang and Rothman 2004). These are important because there is a reduction or elimination of cross-contamination which may cause false positives. qPCR provides an estimate of the number of copies of target DNA which can be used to estimate the number of cells present and allows one to track PCR results as they happen in real time.. qPCR detects both infectious and non-infectious pathogens. These methods are being used for ambient waters and recreational waters but not yet for drinking water. Studies have shown for example that adenoviruses concentrations in environmental samples based on qPCR were higher than cultural method and this difference was assumed to be because of the presence of non-infections target and low sensitivities of culture methods to serotypes present in the environmental samples (He 2004). Thus there is concern that this method does not allow for evaluation of the adequacy of treatment like water disinfection, where the microorganisms are killed but their DNA (or RNA) remains. However for groundwaters and surface waters the methods are beginning to see significant application (Wong et al. 2007) The methods have been successfully used for recreational waters to address viral hazards (Xagorarki et al. 2007)

Source Tracking Tools

Microbial source tracking (MST) assists in the identification of the source of microbial fecal contamination in the aquatic environment (Domingo et al. 2007). Fecal indicator systems *E.coli* and fecal coliform bacteria have been used to address water quality associated with designated uses of swimming, fishing and as a drinking water supply. However, it was recognized early on that although these indicator systems may "indicate" that fecal pollution was present and at what level, the source of the pollution could not be ascertained. Hence, the ability to remediate or control the source was limited to surveys of land use. Point source pollution (e.g., sewage treatment plant discharges) was fairly easily identified but non-point sources (e.g., run-off from agricultural sites with manure, biosolids or septage; wildlife including coastal/aquatic birds as well as septic tanks and even sewage spills and leaks) were extremely difficult to identify. Human sewage, carrying over 100 known pathogens including human specific enteric viruses, had also been traditionally thought to be a greater hazard to human health than animal sources. However, with the rise of zoonotic pathogens such as *E.coli* 0157:H7 and *Cryptosporidium,* the risk of waterborne disease associated with microorganisms originating from animal fecal sources became of equal concern to public health.

As genetic methods advanced, the ability to amplify specific genetic material from bacteria isolated from feces and water in the laboratory using the polymerase chain reaction (PCR) developed. Thus, researchers began using these methods to address the relationship, primarily from a genetic perspective, between animal hosts and their normal gut flora of bacteria, including coliform bacteria, *E.coli* and enterococci. The methods developed were classified into two categories: 1) library-dependent methods, that typically utilize databases containing genotypic or phenotypic "fingerprints" of organisms from known sources to which fingerprints from unknown isolates can be compared; and 2) library-independent methods that define target host-specific microorganisms or genes using various molecular methods, usually by polymerase chain reaction (PCR) (Bernhard and Field 2000a, 2000b; Hsu et al. 1995; Scott et al. 2005, Fields et al. 2003b). The library-independent methods, now referred to as host-specific methods, are now preferred as they cost less and can produce results within 1-2 days, which is significantly better than fingerprinting methods such as ribotyping, which can take 7-14 days. Also, because library-dependent methods rely on building libraries which are specific to a geographic area, they have been found to be less accurate and precise (See J. Water and Health 2003, vol 1. and Water Research, 2007 vol. 41 (11)).

- The evolution of the MST field has shifted from library-dependent methods to library-independent methods targeting host-specific microbes or genetic targets detected and measured by PCR.

Since multiple pathogen analysis is more complex, knowing the source can begin to narrow the search for hazard (Table 7.5)

Table 7.5. Some Host Specific Markers for MST

Marker	Applications	Notes & Limitations
Fluorometry (Hartel et al., 2007) Recommended for evaluation*	Sewage Detection of optical brighteners Can use handheld instrument in the field.	Must look at difference in fluorometric value before and after UV to address level as organics interfere. Waters must be characterized for TOC. Limited evaluation in the field.
Bacteriodes (Bernhard and Fields, 2000a & b; Bernhard et al., 2003; Okabe et al., 2007; Savichtcheva et al., 2007) Recommended†	Non-cultivatable unique targets from the 16s rRNA for human marker cow marker pig marker	Found in 1 ul of sewage diluted /1 liter of water. Levels in animal waste not well evaluated. Stability not well known.
Enterococci esp gene (Scott et al. 2005 Ahmed et al., 2007 Whitman et al. 2007) Recommended†	Cultivatable based method Human Sewage marker	100% + in untreated sewage; 97% + in septic tanks Must examine *E. facieum* specifically. Found in 1% of enterococci population.
Bifidobacterium (Plummer and Long, 2007)	Cultivatable human marker, sorbitol-fermenting anaerobe.	Limited applicability as sensitive to oxygen, as a strict anaerobe.
Rhodococcus cropophilus (Plummer and Long, 2007)	Cultivatable indicative of ruminants.	Not host specific, found in wildlife and domestic ruminants. 50CFU/100ml the level of concern. Not detected in water on a routine basis.
Adenoviruses (Jiang et al., 2006; Xagoraraki et al., 2007) Recommended for evaluation*	PCR can distinguish animal vs human. qPCR method available.	Dependent on excretion rates. Not widely evaluated in animal wastes. Survives better in surface waters as resistant to sunlight.
Enteroviruses (Fong et al., 2005) Recommended for evaluation*	PCR can distinguish animal vs human	Dependent on excretion rates. Not widely evaluated in animal wastes.
Polyomaviruses (Brownell et al., 2007; McQuaig et al. 2006)	Human sewage marker; not cultivatable; found in 80% of humans in urine.	Has not been widely evaluated. Estimated concentrations not known.
Coliphage (Hsu et al., 1995) Recommended for evaluation*	Cultivatable, can sort human, from animals from birds.	Based on genetic type however, there is overlap.

* Recommended that these markers receive more evaluation for distinguishing sources and for applicability.

† Recommended that these markers be utilized in watershed studies

7.4 PLANNING A LARGE SCALE WATERBASIN MONITORING PROGRAM

7.4.1 Background

The most significant problems with the current monitoring programs for addressing microbial water quality and thus safety are a completely inadequate data base which can not be used to quantify and prioritize the microbial risks and thus the control of the pollution problems. Beaches continue to be closed, new pathogens continue to cause not just disease by waterborne outbreaks and estimates of endemic waterborne disease and mortality continue to be in the millions for both the developed and developing regions of the world, respectively. Obstacles to adequate data collection include the lack of a planned study that can incorporate local data into large scale assessment, use and interpretation of new tools and techniques, and lack of political will which denotes the perceived lack of usefulness of the monitoring data.

Most large scale basin assessments have focused on the protection of drinking water uses. Toward this end, the World Health Organization has developed the *Water Safety Plans* which follow much of the HACCP concepts discussed in Chapter 6 (Deere and Davison). Microbial pathogens represent the most consistent risk to drinking and recreational water safety. For drinking water, the prevention and reduction of microbial risks has been presumed to be achieved via a multiple barrier approach (watershed protection, water treatment including filtration and disinfection, and protection of the integrity of the distribution system) which has been advocated by public health professionals for more than a century and is considered best practice. However, watershed protection is often just given lip service as the geohydro-political boundaries and development of policies governing protection are complex. In theory the assessment involves understanding the hazards and risks, and establishing appropriate barriers to mitigate or prevent those risks.

Water Safety Plans (WSP) have three components which include system assessment, identification of control measures and finally development of a management plan. Like the HACCP system, this is to follow the drinking water supply chain from the source to the tap providing an evaluation of the risks, and assurances of safety. Water quality objectives associated with public health targets are the ultimate goals

A new approach is needed which takes advantage of current monitoring data, new techniques and articulation of the health targets at large scales and so that various government agencies can support the program financially and with human resources. The principals of a monitoring program should include the following.

1. Work toward the implementation of WSP/HACCP

2. Implement at local scales after a large-scale assessment is undertaken focusing on the water balance, land-use, hydrodynamics and trophic levels.
3. Delineate hotspots (Areas of Concern) including the identification of sources, exposures, reservoirs (sediments).
4. Use the Best-available technology, new tools and approaches for quantifying microbial pathogens, tracking sources, modelling and georeferencing.
5. Include the various stakeholders with the scientific expertise needed to address
   a. The plan
   b. The locations
   c. The time frame
   d. The methods and laboratories
   e. The data summary (geo referencing)
   f. The information sharing
   g. The public assurances

7.4.2. Steps Associated with Large Scale Water Quality Monitoring Programs

All monitoring programs should begin with a discussion regarding the purpose and application of the information and data. Some microbial water quality data may already be available but at various scales that may not meet the needs for addressing risk or decision making. Thus the information gaps that are identified both in space and time will define the tools and techniques that will be used and the necessary scientific expertise, laboratory and human resources needs. (Figure 7.2). This will also help to define the necessary political and government agencies that need to be involved. For microorganisms as already stated the purpose is often to address problem exposures sites (eg beaches) at a very local level but programs should be focused on assessment and hopefully lead toward investment to prevent the level of microbial pollution associated with significant health risks. Scale is extremely important for the sampling of microorganisms as samples are collected at the field (macro and micro scales) changing in time as quickly as minutes, but many of the international water basins and the hydrodynamics of the system are at kilometer scales with a time range of days to years (Table 7.6). Thus data sets are needed that can begin to address the local within the larger basin with a nested approach.

Table 7.6. Spatial and Temporal Scale Considerations Important for Microbial Water Quality

Earth	Continents /Countries Water basins	Lakes and Rivers	Wastewater discharge /Agricultural field
Spatial Scales:			
>10,000 kilometers	1 to 100 kilometers	100 meters to kilometers	centimeters to meters
Temporal Scales:			
Years to Decadal to 100 Years or longer (eg. ENSO)	Years	Daily to Years	Minutes to Daily

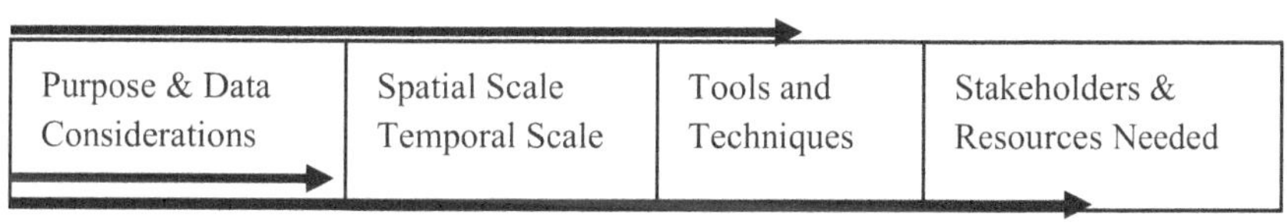

Figure 7.2. Pathway of Thinking for Planning an Assessment of Microbial Water Quality

In addition tools and techniques that have not traditionally been used for microbial water quality data sets need to be integrated into the planning. This includes:

- Use of geographic information systems for mapping, inventory and spatial modeling.
- Use of remote sensing to address scale, land use, climate, and trophic status of waters.
- Use of hydrodynamic modeling for transport, and integration of data at various spatial and temporal scales.
- In-situ automated data acquisition including precipitation, flows, physical/chemical water quality information.

The first step should begin by defining the basin at the largest scale and the water resources which includes in every system the available ground waters and aquifers as well as the "engineered hydrologic cycle"(dams, controlled releases etc). For microorganisms there will be interest in the sources that contribute to the pollution and the exposure sites (water supplies and beaches). Because the Global Ocean Observing System will be in place and satellite and remote sensing programs are also becoming global (See Chapter 5, Malone and Rockwell) these should be used for all international basins and governments should invest in the coordination and support of such systems along with data sharing. The understanding of land use and climate as these contribute to nutrients and trophic status will not only be useful in examining the risk of hazardous algal blooms but in understanding on a larger scale the sources of pathogens and microbial risks.

Trophic status can be generally assessed with the surrogate chlorophyll a, and recently studies using Land Sat have been able to ground truth models which could be used to examine on a large scale the spatialtemporal dynamics of water quality (Wiangwang et al. 2006; Zhang et al. 2007; Qi 2007) Thus it is recommended that a global effort be made to understand eutrophication and water quality risks of the large water basins of the world.

STEP 1. Defining the Water Basin

The Purpose

1. Delineate the water basin (surface and groundwater).
2. Address high risk areas (inputs; high concentrations of microbes)
3. Identify high exposure areas

Data Considerations

4. Engineered hydrologic cycle
5. Land use
6. Animal and human populations and their wastes
7. Climate (precipitation, temperature, winds, relative humidity)
8. Trophic status

Spatial and Temporal Scale

9. Largest scale (identify number of geopolitical layers)
10. Seasonal multi-year assessment

Tools and Techniques

11. Remote Sensing
12. Land Sat
13. Rain and river gauges
14. In-situ platforms (weather, ocean, lake systems)

Stakeholders and Resources

15. Hydrologists, engineers, climatologists, land use experts, phycologists, social scientists, political leaders; Computational, remote sensing facilities.

Steps 2 and 3 will begin to address water quality at more local scales associated with defining the hazards and the exposures. These should tie into the health goals in the Water Safety Plans as well as hazards and critical control points as described in Chapter 6 (Deere and Davison). Step 2 will focus on characterization of the fecal loading, pathogen occurrence associated with various sources. Land use patterns and climate should be addressed. The first goal would be to map the current *E.coli* data sets. Generally, wastewater facilities do not address *E.coli* or alternative indicators, thus for these sources it is recommended that pathogen monitoring be undertaken, in addition to *E.coli*.

Wastewater monitoring addresses the health of the community (representing what infections the population are experiencing) as well as monitors the inputs into the basin. Each group would coordinate their data acquisition using georeferencing and a common plan, using consistent techniques and tools, thus creating "nested" data sets that will feed into the large basin assessment.

STEP 2. Hazard identification

The Purpose

1. Define microbial water quality using global yard stick *E.coli* (surface and groundwater)
2. Address the relative contributions of sources (human and animals)
3. Identify the prevalence of pathogens of concern

Data Considerations

4. Fecal indicators in water and sediments
5. Mapping occurrence
6. Hot spots and control sites (for HACCP)
7. Seasonal monitoring data (storms, algal blooms)
8. Prevalence and concentrations of pathogens

Spatial and Temporal Scale

9. Discharges from point and non-point sources impacted areas
10. Seasonal multi-year assessment

Tools and Techniques

11. *E.coli* monitoring
12. Alternative indicators
13. Use of source tracking markers
14. Algal toxin methods
15. Pathogen assessment
16. qPCR for pathogens
17. Geo-referencing/ GIS

Stakeholders and Resources

18. Microbiologists, phycologists, public health and water supply professionals, local leaders. Environmental microbiology and molecular biology laboratories.

Step 3 is particularly focused on the transport from the sources to the exposure sites. Very little work has been done at the larger scale in regard to natural barriers (e.g. wetlands, soils) and reservoirs of microbial fecal indicators and pathogens, including wildlife and sediments (Karim et al., 2004). Understanding survival is also key to this part of the assessment. Temperature plays a key role in the inactivation of micororoganisms but it is now clear that sunlight inactivation in surface waters is extremely important and will need to be assessed (Liu et al., 2006). The over- wintering of pathogens and seasonal considerations has not been adequately described in most studies. The integration of hydrodynamic modeling and microbiological data sets are critical.

STEP 3. Exposure Assessment

The Purpose

1. Define microbial pathogen risks at recreational and drinking water sources

Data Considerations

2. Fecal indicator quantification from sources to exposure sites
3. Mapping occurrence of pathogens and source tracking markers
4. Relationships between hot spots in time and space
5. Reservoirs of indicators and pathogens (sediments and wildlife)
6. Survival
7. Growth dynamics for algae
8. Identify control sites (for HACCP)

Spatial and Temporal Scale

9. Sample exposure sites: water supplies, beaches, ground waters
10. Transect and nested data from source to exposure sites address, travel time
11. Seasonal multi-year assessment

Tools and Techniques

12. Hydrodynamic models at various scales
13. Models of indicator transport
14. Use of attenuation factors in models
15. In-situ platforms support for modeling
16. Examination of predictive models for blooms and contamination
17. Pathogen monitoring validation for models
18. QMRA to address probability of infection

Stakeholders and Resources

19. Microbiologists, phycologists, hydrologists, public health and water supply professionals, local leaders. Environmental microbiology and molecular biology laboratories. Computational and hydrodynmic laboratories.

STEP 4. Management and Control Assessment

The Purpose

1. Define water quality, safety targets and goals for the water basin

Data Considerations

2. Examine reductions by engineered and natural systems
3. Evaluate the reliance on dilution and attenuation to achieve safety
4. Risk of basin in regard to extreme weather events and climate change
5. Assurance regarding control sites (for HACCP)

Spatial and Temporal Scale

6. Identify key critical control sites for addressing baseline

7. Over time, address natural barriers and engineered systems
8. Reliability, variability and risk to extreme events or failures

Tools and Techniques

9. On-line process control
10. Large volume high through put indicators
11. Biological verification controls (seeded)
12. Statistical failure assessment
13. Pathogen monitoring validation of the ZERO
14. QMRA

Stakeholders and Resources

15. Mangers, microbiologists, engineers, water professionals and local leaders. Environmental microbiology and molecular biology laboratories. Pilot scale systems for validation.

The Monitoring Plan at the various scales will be key to successful management at the larger scale but must include the establishment of baseline conditions, the measuring of implementation of changes, as well as determining the effectiveness of such implemented changes in obtaining the desired outcomes of environmental improvements. Thus, this "nested monitoring" approach--where monitoring of local indicators is connected to larger scale monitoring at the regional scale will allow for a basin assessment and improved prioritization and strategic investment. From the perspective of public health, the public assurances are critical to gaining the support of the citizenry.

7.5 CONCLUDING REMARKS

The largest microbiological monitoring program of an international basin took place in 1914 in the Great Lakes basin and to date no other study of this stature has been implemented. There are a number of large international and national monitoring programs, but the purposes are single minded, the data are gathered in a fragmented fashion and very little spatial and temporal analysis is undertaken with the data sets. The issues of developing baselines and large scale information across numerous watersheds or aquifers are complicated but it is clear that goals, methods, data analysis and information sharing are critical elements. All programs should demonstrate inclusion of some of the key principles, georeferencing, and improved spatial and temporal analysis within the key water basins.

Evaluation of water quality data and climate factors, in particular may need to use statistical techniques that examine the larger drainage basins at monthly temporal scales. Studies on 50 years of waterborne outbreaks in the US, for example, found that in both groundwater and surface waters, relationships in time (associated with precipitation) and space (key watersheds at risk) could be determined but only at larger scales and times associated with months, not days. (Rose et al. 2000; Currieo et al. 2001).

Below are described briefly two large scale microbiological monitoring programs and finally, recommendations for improving microbiological water quality assessment.

7.5.1 The European Blue Flag Campaign

Fresh and marine recreational waters are monitored for fecal microbial indicators around the world. The European Union has 14,345 coastal beaches and 6,749 freshwater bathing areas and has implemented a EU Bathing Water directive which addresses bacteriological monitoring for the purpose of classification of potential recreational waterborne disease risk. Thus if one were to examine at the largest scale, the coastline is being monitored, but analysis is done most often at the country level focused on the associated beach designations and binning of the data into categories and application of a variety of tools and methods. For example, Havemister et al. (1991) examined in 1 year 222 samples at beaches in the Northsea, Baltic Sea and inland waters for the purpose of evaluating methods; in Ireland three years of monitoring (1999-2001) were undertaken to address the classification of beaches (Chawla et al. 2005) and finally in Spain, in 1998, monitoring of 15 beaches of Gipuzkoa (the Basque Country, Spain) was implemented to address the application of an alternative fecal indicator, the coliphage (Ibarluzea et al. 2007).

The proposed EU Directive (Excellent, Good, Poor) uses enterococci for the ranking and an assessment of 'Excellent' classification is required at present for the Blue Flag award. This program has led to:

- Consistent method for monitoring
- Approach for interpreting the results
- Program that is implemented over a large scale at an international level

7.5.2 The US Information Collection Rule

The *Information Colletion Rule* was developed in order to address the safety of drinking water surface supplies in the US from the emerging parasite *Cryptosporidium* and balance the risks associated with the treatment of microorganisms particularly by disinfection and the development of potentially carcinogenic disinfectant by-products (USEPA 2003). The goal was to develop a national estimate of occurrence of microbial pathogens in drinking water sources, examine the necessary treatment for reduction of the risk. While *E.coli*, viruses and the parasites *Cryptosporidium* and *Giardia* were monitored for, it was the chlorine-resistant *Cryptosporidium* that drove the analysis, As a part of this rule the following was accomplished:

- A standard method for the parasites and viruses as well as a laboratory certification procedure were put into place to provide a basis for development of a national resource of laboratories.
- Monthly monitoring took place over 18 months to capture annual fluctuations (July 1997 to December 1998 and a supplemental took place from March 1999 to February 2000).
- The intakes to drinking water facilities 47 large systems (>100,000 people) and 40 medium systems (serving 10,000 to 100,000 people) were monitored.

Due to the recovery issues the *Cryptosporidium* data were adjusted for the mean by the analytical method recovery of 12%. Of the 350 samples 56% had detectable oocysts, and the median and 90th percentile estimates of *Cryptosporidium* levels in US national waters during this time frame were 0.048 and 1.3 oocysts/L, respectively. These data were used to formulate the reduction goals for treatment associated with the level of contamination in the source water. In highly contaminated waters, sand filtration is not seen as providing a significant enough barrier and thus utilities will need to consider membrane filtration or UV disinfection. These data have also encouraged drinking water utilities to look up stream at sources of oocysts entering their waters and ways to protect the watershed.

7.5.3 Recommendations for the Implementation of Microbiological Monitoring of Transboundary Waters

The many facets of globalization increasingly influence the world we live in, impacting either directly or indirectly the risk of water pollution and waterborne disease. Both the safety of drinking water and recreational waters drive the need for an improved understanding of microbial quality of the water. Due to the nature of microbial pollution, emerging microbial risks, and exposure at the community level including the sensitive populations a shift in regard to the monitoring and management of transboundary waters is needed. Thus;

- Water Safety Plans should be expanded to recreational waters and include assessment of return flows in addition to protection of the water supply at the large water basin scale.
- QMRA should be used to help set health targets, along with pathogen specific, disease specific goals.
- Remote sensing, source tracking methods, qPCR for pathogens should be used strategically.
- Information should be gathered and analyzed from a spatial and temporal perspective.

- Stakeholders should collectively share the costs and benefits of planning, implementation and knowledge building to improve decisions in the management of transboundary waters for public health protection.

Water quality remains the most important and critical component of the safety and security for development of any community, both from a public health and economic standpoint. Controlling the microbiological threat to water has been part of human history and sanitary science was an essential development in the protection of public health. This type of scientific endeavor has been devalued and we have focused on the individual rather than on communities. We have failed to utilize all the new tools, approaches and technologies available to us to address water quality and health issues. We must set targets, we must be able to measure and interpret that information for the betterment of our earth. Access to information and knowledge is a power that will move us toward improved prioritization and investment. It is our collective responsibility to protect water resources for the future and for all who share the resource.

7.6 REERENCES

Ahmed, W. Stewart, J., Gardner, T., Powell, D., Brooks, P. Sullivan, D. and Tindale, N. Sourcing feacal pollution: A combination of library-dependent and library-independent methods to identify human faecal pollution in non-sewered catchments. Water Res. Vol.41: 3771-1779.

Ashbolt, N.J., Grabow, W.O.K. and Snozzi, M. 2001. Indicators of Microbial Water Quality. *Water Quality: Guidelines, Standards and Health*, L. Fewtrell and J. Bartram (Eds.). IWA Publishing, London, UK, 289–315.

Beck, R. W. 2000. *A chronology of Microbiology*. ASM Press, Washington DC.

Bernhard, A.E. and Field, K.G., 2000a. Identification of nonpoint sources of fecal pollution in coastal waters by using host-specific 16S ribosomal DNA genetic markers from fecal anaerobes. Appl. Environ. Microbiol. 66 (4), 1587-1594.

Bernhard, A.E., and Field, K.G., 2000b. A PCR assay to discriminate human and ruminant feces on the basis of host differences in *Bacteroides-prevotella* genes encoding 16S rRNA. Appl. Environ. Microbiol. 66 (10): 4571-4574.

Bernhard, A. E., T. Goyard, M. T. Simonich, and K. G. Field. 2003. Application of a rapid method for identifying fecal pollution sources in a multi-use estuary. Water Res 37:909-13.

Brownell, M.J., Harwood, V.J., Kurz, R.C., McQuaig, S.M., Lukasik, J., and Scott, T.M. 2007.Confirmation of putative stormwater impact on water quality at a Florida beach by microbial source tracking methods and structure of indicator organism populations. Water Res. 41:3747-3757.

Bruneau, A. H. Rodrigue, J. Ismael, R. Dion, and R. Allard. 2004. Outbreak of E. coli O157:H7 associated with bathing at a public beach in the Montreal-Centre region. Can. Commun Dis Rep. 30(15): 133-136.

Cabelli, V.J. (1978) Obligate anaerobic bacterial indicators. In Indicators of Viruses in Water and Food (ed. G. Berg), pp. 171–200, Ann Arbor Science, Ann Arbor, MI.

Chawla R, Real K, Masterson B. 2005 An assessment of the impact of the proposed EU bathing water directive on Irish coastal bathing area compliance.*Water Sci Technol.* 51(3-4):225-30.

Curriero, F.C., Patz, J.A., Rose, J.R., Lele, S. 2001. The Association Between Extreme Precipitation and Waterborne Disease Outbreaks in the United States, 1948-1994. *Amer J. Pub. Hlth.* 91 (8). 1194-1199.

Davies,C.M., Long, J.A.H., Donald, M. and N.J. Ashbolt. 1995. Survival of Fecal Microorganisms in Marine and Freshwater Sediments. *Appl. Environ. Microbiol.* **61**(5): 1888-1896

Domingo, J.W.S., Bambic, D.G., Edge, T.A., Wuertz, S. (2007) Quo vadis source tracking? Towards a strategic framework for environmental monitoring of fecal pollution. Water research 41: 3539 – 3992

Eisenstein, B.I. 1990. The polymerase chain reaction: a new method of using molecular genetics for medical diagnosis. *New England Journalof Medicine* 322, 178–183.

Field, K. G., A. E. Bernhard, and T. J. Brodeur. 2003a. Molecular approaches to microbiological monitoring: fecal source detection. Environ Monit Assess 81:313-26.

Field, K.G., E.C. Chern, L.K. Dick, J.A. Fuhrman, J.F. Griffith, P.A. Holden, M.G. LaMontagne, J. Le, B.H. Olson, and M.T. Simonich. 2003b. A comparative study of culture-independent, library-independent genotypic methods of fecal source tracking. J. Water and Health 1: 181-194.

Fong, T.T., Griffin, D.W., Lipp, E.K., 2005. Molecular assays for targeting human and bovine enteric viruses in coastal waters and their application for library-independent source tracking. Appl. Environ. Microbiol. 71 (4), 2070-2078.

Gleeson, C. and Gray, N. 1997. The Coliform Index and Waterborne Desease. E and FN Spon, London.

Haas, C. H., Rose, J.B, and Gerba, C.P.. (1999). *Quantitative Microbial Risk Assessment.* New York, NY, John Wiley and Sons.

Harrison, S. and S. Kinra. 2004. Outbreak of Escherichia coli O157 associated with a busy bathing beach. *Commun Dis Public Health.* **7**(1): 47-50

Hartel, P.G., Hagedorn, C., McDonald, J.L., Fisher, J.A., Saluta, M.A., Dickerson Jr.,J.W., Gentit, L.C., Smith, S.L., Mantripragada, N.S., Ritter, K.J., and Belcher, C.N. 2007 Exposing water samples to ultraviolet light improves flurormetry for detecting human fecal contamination. Water Res. 41: 3629-3642.

Havemeister G, Aleksic S, Bockemühl J, Heinemeyer EA, Müller HE, Von Pritzbuer E. 1991 Comparative studies of fresh and seawater for the determination of total coliform and fecal coliform bacteria according to the European Economic Community guideline 76/160 (bathing water) by the use of the most-probable-number method with BRILA-MUG broth and differentiation according to the drinking water ordinance] *Zentralbl Hyg Umweltmed.* May;191(5-6):523-38.

He J.W., and Jiang, S. (2005). Quantification of Enterococci and Human Adenoviruses in Environmental Samples by Real-Time PCR. Appl Environ Microbiol. **71**(5): 2250-2255

Hsu, F.C., Sheieh, Y.S.C., van Duin, J., Beekwilder, M.J., and Sobsey, M.D., 1995. Genotyping male-specific RNA Coliphages by hycridization with oligonucleiotide probes. Appl. Environ. Microbiol. 61 (11): 3960-3966.

Howard-Jones, N. 1973, Gelsenkirchen Typhoid Epidemic of 1901, Robert Koch, and the Dead Hand of Max von Pettenkofer, British Medical Journal, 1973, 1, 103-105

Hrudey,S.E. and EJ Hrudey 2004. Safe Drinking Water *Lessons from Recent Outbreaks in Affluent Nations* IWA, London, England ISBN: 1843390426

Ibarluzea JM, Santa Marina L, Moreno B, Serrano E, Larburu K, Maiztegi MJ, Yarzabal A. 2007 Somatic coliphages and bacterial indicators of bathing water quality in the beaches of Gipuzkoa, Spain.*J. Water and Health* Sep;5(3):417-26.

Jiang, S. C. (2006). "Human adenoviruses in water: occurrence and health implications: a critical review." Environ. Sci. Technol. **40**(23): 7132-7140.

Karim MR, Manshadi FD, Karpiscak MM, Gerba CP. 2004. The persistence and removal of enteric pathogens in constructed wetlands. Water Res. 38(7):1831-1837

Levy, DA., Bens, M.S, Craun, G.F., Calderon, R.L, and B.L. Herwaldt. 1998. Surveillance for waterborne-disease outbreaks-United States, 1995-1996. *MMWR.* **46** (SS-5): 1-34.

Liu, L., Manthas, P., Molloy, S.L., Whitman, R.L., Shiverly, D.A., Nevers, M., Schwab, D.J. and Rose, J.B. 2006. Modeling The Transport And Inactivation Of *E. Coli* And Enterococci In The Near-Shore Region Of Lake Michigan. *Environ. Sci and Tech.* 40(16):5022-5028.

Maier, R.M, I.L. Pepper, and C.P.Gerba. 2000. Environmental Microbiology. (Chapter 20-Indicator Microorganisms page 491) Academic Press, San Diego

Medema, G., Teunis, P., Blokker, M., Deere, D., Davison, A., Charlis, P. and Loret, J. 2006. WHO Guidelines for Drinking Water Quality *Cryptosporidium*, EHC *Cryptosporidium* Draft

McQuaig, S.M., Scott, T.M. Lukasik, J., and Farrah, S.R. 2006. Detection of human-derived fecal pollution in environmental waters by use of a PCR-based human polyomavirus assay. Appl.Environ.Microbiol. 72 (12)7567-7574.

Mullis KB, Faloona FA. 1987. Specific synthesis of DNA in vitro via a polymerase-catalyzed reaction. *Methods Enzymol*; 155: 335–50.

Mullis KB. 1990. The unusual origin of the polymerase chain reaction. *Sci Am*; 262: 56–65.

Noble, R.T., S.M. Allen, A.D. Blackwood, W. Chu, S.C. .Jiang, G.L. Lovelace, M.D. Sobsey, J.R. Stewart, and D.A. Wait. 2003. Use of viral pathogens and indicators to differentiate between human and non-human fecal contamination in a microbial source tracking comparison study. *J. Wat. Health.* 1(4): 195-208

Okabe, S, Okayama N, Savichtcheva O, Ito T. 2007. Quantification of host-specific Bacteroides-Prevotella 16S rRNA genetic markers for assessment of fecal pollution in freshwater. *Appl Microbiol Biotechnol.* 2007 Mar;74(4):890-901. Epub 2006 Dec 1.

Paunio, M. R. Pebody, M. Keskimaki, M. Kokki, P. Ruutu, S. Oinonen, V. Vuotari, A. Siitonen, E. Lahti, and P. Leinikki. 1999. Swimming-associated outbreak of Escherichia coli O157:H7. *Epidemiol Infect.* **122**(1): 1-5

Pepper, I. L., Gerba, C.P. and M.L Brusseau (Eds). 2006 Environmental and Pollution Science. Boston: Academic Press

Persing DH. 1993. Diagnostic molecular microbiology: current challenges and future directions. *Diagn Microbiol Infect Dis*; 16: 159–63.

Plummer, J.D. and Long, S.C. 2007. Monitoring source water for microbial contamination: Evaluation of water quality measures. Water Res. Vol.41: 3716-3728.

Qi, J., 2007 Remote sensing's role in water quality assessment, Water 21, February 07, pg 31

Regli S, Rose JB, Haas CN, and Gerba CP. 1991. Modeling the Risk from *Giardia* and Viruses in Drinking Water. *J. Amer. Water Works Assoc.* Nov. 83:76-84.

Rose JB, Haas CN, and Regli S. 1991. Risk Assessment and Control of Waterborne Giardiasis, *Amer. J. Public Hlth Assoc.* 81(6): 709-713.

Rose, J.B., Daeschner, S., Deasterling, D.R., Curriero, F.C., Lele, S. and Patz, J. 2000. Climate and waterborne disease outbreaks. *J.Amer. Water Works. Assoc.* 92:77-87.

Rose, J.B. and Farrah, S.R. 2007. Microbial health risks and Water Quality IN:*Urban Water Security: Managing Risks*, Blanca Jimenez and Joan Rose (EDS) UNESCO

Savichtcheva, O., Okayama, N. and Okabe, S. 2007. Relationships between *Bacteroides* 16s rRNA genetic markers and presence of bacterial enteric pathogens and conventional fecal indicators. Water Res. Vol.41: 3615-3628

Scott, T.M., Jenkins, T.M., Lukasik, J., Rose, J.B., 2005. Potential use of a host associated molecular marker in Enterococcus faecium as an index of human fecal pollution. Environ. Sci. Technol. 39(1), 283-287.

USEPA, 2003. http://www.epa.gov/EPA-WATER/2003/August/Day-11/w18295a.htm

National Primary Drinking Water Regulations: Long Term 2 Enhanced Surface Water Treatment Rule [Federal Register: August 11, 2003 (Volume 68, Number 154)]

White TJ, Madej R, Persing DH. 1992The polymerase chain reaction: clinical applications. *Adv Clin Chem*; 29: 161–96.

Whitman, R.L. Przybyla-Kelly, K, Shively, D.A. and Byappanahalli, M.N.2007. Incidence of the enterococcal Surface Protein (esp) Gene in Human and animal Fecal Sources. Environ. Sci & Tech. 41: 6090-6095.

Wiangwang, N., Messina, J.P., Qi, J., and D.L. Skole. 2006. Water Clarity/ Trophic Condition Assessment of the Bight of Bangkok using Satellite Remote Sensing Imagery. International Journal of Geoinformatics. 2(3) 43-50.

Wolman, A. and Gorman, A. E. 1931. *The significance of waterborne Typhoid Fever outbreaks, 1920-1930.* William and Wilkins Company, Baltimore.

World Health Organization, 1999 *Health-Based Monitoring of Recreational Water: The Feasibility of a New Approach (The 'Annapolis Protocol').* 1999. (WHO/SDE/WSH/99.1). World Health Organization – Sustainable Development and Healthy Environments.

Xagoraraki, I.. Kuo, D.H-W; Wong,K., Wong, M. and Rose, J.B. 2007. Occurrence of Human Adenoviruses in Two Great Lake Recreational Beaches. *Applied Environ Microbiol In press*

Yang, S. and Rothman R. 2004. PCR-based diagnostics for infectious diseases: uses, limitations, and future applications in acute-care settings. *The Lancet Infectious Diseases*, 4 (6), 337-348

Zhang, J., Hu, F., Qi, J., Torbick, N., and Zhang, H. 2007. Mapping Chlororphyll-a Concentration in West Lake China using Landsat 7 ETM+. *J. Great Lakes Res.* Accepted

Index

Numbers in italics indicate figures.

IWA Publishing's authorised EU representative for General Product Safety Regulations is Diane D'Arras, 15 rue Duret, 75116 Paris, France, e-mail: safety@iwap.co.uk.

Printed and bound by CPI Group (UK) Ltd, Croydon, CR0 4YY
16/06/2026
02140353-0001